H. G. Bronn, Ernst Haeckel, and the Origins of German Darwinism

Transformations: Studies in the History of Science and Technology
Jed Z. Buchwald, general editor

H. G. Bronn, Ernst Haeckel, and the Origins of German Darwinism

A Study in Translation and Transformation

Sander Gliboff

The MIT Press
Cambridge, Massachusetts
London, England

For information about special quantity discounts, please email special_sales@ mitpress.mit.edu.

This book was set in Sabon by SPi, Puducherry, India, and was printed and bound in the United States of America.

Library of Congress Cataloging-in-Publication Data

Gliboff, Sander.
H. G. Bronn, Ernst Haeckel, and the Origins of German Darwinism : A Study in Translation and Transformation / by Sander Gliboff.
p. cm.—(Transformations : studies in the history of science and technology)
Includes bibliographical references and index.
ISBN 978-0-262-07293-9 (hardcover : alk. paper)
1. Evolution (Biology)—Germany—History. 2. Bronn, H. G. (Heinrich Georg), 1800–1862. 3. Haeckel, Ernst Heinrich Philipp August, 1834–1919.
I. Title.
QH366.2.G55 2007
576.80943—dc22

2007039859

10   9   8   7   6   5   4   3   2   1

To Renate and Hannah

Contents

Acknowledgments

This book grew out of my graduate work at Johns Hopkins on Ernst Haeckel and Paul Kammerer. My thanks to Sharon Kingsland for her advice and patient reading of drafts; to Dan Todes, whose work on Russian Darwinism was an early model for me; and to Bob Kargon for getting me interested in scientific translation. Further intellectual and moral support came from fellow graduate students Lloyd Ackert, Keith Barbera, Kathleen Crowther, Greg Downy, Carl-Henry Geschwind, Hunter Heyck, Tom Lassman, Josh Levens, David Munns, Buhm Soon Park, and Shahana Sarkar. My work was also aided by a dissertation improvement grant from the National Science Foundation and a foreign travel grant from the Zanvyl Krieger School of Arts and Sciences.

My project expanded to include Bronn, when I was a postdoctoral fellow at Northwestern University, in the program on Science in Human Culture. My thanks to Ken Alder and David Hull for their advice and support while I was there.

My first findings on Bronn appeared in the *Journal of the History of Biology* 40 (2007), under the title "H. G. Bronn and the History of Nature." Material from that article is reused here with kind permission of Springer Science and Business Media.

At Indiana University, my ideas about Haeckel, Bronn, and Darwin coalesced into this book. Thanks are due to my IU colleagues Jim Capshew, Bill Newman, and Jutta Schickore, as well as to participants in our Biology Studies Reading Group, HPS Reading Group, and my graduate seminars on Darwin, who read various excerpts from this manuscript, including: Colin Allen, Nico Bertoloni Meli, Jordi Cat, Fred Churchill, Lisa Lloyd, Jason Byron, Matt Dunn, Melinda Fagan, Stephen Friesen, Joel Klein, Costas Mannouris, Evan Ragland, Sean Valles, and others. Special thanks to Anne Mylott, for reading and commenting on an entire draft. My work also benefited from a Faculty Summer Fellowship from the Office of the Vice President for Research, and a Faculty Fellowship from the College Arts and Humanities Institute.

Thanks also to Patricia Princehouse, Judy Schloegel, and Jim Strick.

Last, but not least, my wife Renate Kasak and our daughter Hannah Kasak-Gliboff have put up with this project for a long time. I was already working on it when Hannah learned to say "*Bu*," an early utterance of which Renate and I were proud, even though we weren't sure what it meant or even what language it was in. We had to judge by the context in which it was spoken and the intellectual concerns and commitments of the speaker. For sharing such important exercises in translation and interpretation with me, and much else besides, I dedicate this book to Renate and Hannah: *Dieses Bu* is for you.

Introduction

Charles Darwin was notoriously reticent about publicizing his theory of evolution. He had conceived it in broad outline by the late 1830s, drafted his argument concisely in 1842 and at greater length in 1844, but his book *On the Origin of Species* did not see the light of day until 1859. Historians were once perplexed by the seemingly unnecessary and irrational "long delay,"[1] but increasingly they appreciate Darwin's constructive use of the extra time to approach individual scientists with his ideas, test the waters, and develop a network of future supporters. The latest Darwin biographies present Darwin's voluminous correspondence as vivid evidence of his networking skills and his ability to muster broad intellectual, social, and material support for his work from every stratum of Victorian society.[2]

Little note has been taken, however, of how exclusively British his network was. Except for his cultivation of an American ally in the botanist Asa Gray, Darwin left the international reception of his theory largely to chance. Considering the prominence of German biologists in the nineteenth century, Darwin's overtures toward them seem especially weak. He made hardly any until after 1859, when he sent complimentary copies of the *Origin*, out of the blue, to perhaps a dozen German scientists. Even then, Darwin needed help from Thomas H. Huxley to figure out who in Germany would be appropriate recipients.[3]

It is no wonder, then, that Darwin was surprised and gratified when Heinrich Georg Bronn, Germany's most prominent paleontologist, responded positively to receipt of his copy. He promptly reviewed Darwin's book in the leading geological journal under his editorship, and wrote to inquire about translating it into German. He even lined up a publisher for it. Darwin and Huxley had been speculating on help with a German edition from the Swiss morphologist Albert Kölliker, and had not reckoned with such strong interest from Bronn. Darwin wrote to Huxley for advice: "I have had this morning a letter from old Bronn (who, to my astonishment, seems slightly staggered by Natural Selection), and he says a publisher in Stuttgart is willing to publish a translation, and that he, Bronn, will to a certain extent superintend. Have you written to Kölliker? If not, perhaps I had better close with this proposal—what do you think?"[4]

Darwin decided to follow up with Bronn and let him go ahead with the translation:

Dear and much honoured Sir,

I thank you sincerely for your most kind letter; I feared that you would much disapprove of the 'Origin,' and I sent it to you merely as a mark of my sincere respect. . . . I thank you cordially for the notice in the 'Neues Jahrbuch für Mineralogie,' and still more for speaking to Schweitzerbart [sic] about a translation; for I am most anxious that the great and intellectual German people should know something about my book.[5]

Not wishing to appear to expect Bronn to translate the book himself, Darwin suggested he supervise or edit the translation: "I have told my publisher to send immediately a copy of the *new* edition to Schweitzerbart, and I have written to Schweitzerbart that I gave up all right to profit for myself, so that I hope a translation will appear. I fear that the book will be difficult to translate, and if you could advise Schweitzerbart about a *good* translator, it would be of very great service. Still more, if you would run your eye over the more difficult parts of the translation; but this is too great a favour to expect. I feel sure it will be difficult to translate, from being so much condensed." Darwin even suggested that Bronn could help generate greater interest in the translation, and make more money from it, by contributing editorial comments: "How interesting you could make the work by *editing* (I do not mean translating) the work, and appending notes of *refutation* or confirmation. The book has sold so very largely in England, that an editor would, I think, make profit by the translation" (emphases in original).[6]

Darwin should not have been surprised at either Bronn's keen interest in the *Origin* or his open-mindedness about evolution, for during Darwin's "Long Delay," Bronn's research had paralleled Darwin's in some uncanny ways, as he, too, sought alternatives to idealized archetypes, linear scales of development, and a static view of nature. The two naturalists, in their separate ways, both pursued a historical view of life, and both believed that clues to the nature and causes of historical change would emerge not only from paleontology and comparative morphology, but also from studies of variation, geographic distribution, organism-environment interactions, and even domesticated plants and animals and artificial breeding. In the 1840s they each took critical looks at the available theories of organic history. They considered and rejected natural theology, the catastrophism and successive mass extinctions and creations of Louis Agassiz, and Lamarck's theory of species transformation. They were both sympathetic, up to a point, to a variety of successional theories involving staggered sequences of creations and extinctions, leading to gradual turnover of species, but they were also critical of them as they were developed prior to 1840.[7] Both reasoned that the causes of organic change were ultimately to be sought in the changing external environment, rather than any internal law of development or the realm of transcendental archetypes or unfolding plans of nature. Both took it as self-evident that maladapted forms, whether whole species, local varieties, hybrids, or anomalous sports of nature—or, in Darwin's case, even minor individual variations—could not readily perpetuate themselves. And at times in the 1840s, Darwin and Bronn were even consulting each other's work. In Bronn's magnum opus, the multivolume *Handbuch einer Geschichte der Natur* (Handbook of a history of nature; hereinafter, *Geschichte der Natur*),[8] Darwin was one of Bronn's authorities on biogeography and uniformitarian geology. Conversely, Bronn's *Geschichte der Natur* was one of the most heavily annotated works in Darwin's library.[9]

In the late 1850s Bronn and Darwin, separately, came out with their mature theories of organic history and drew international attention and acclaim for them. Bronn had further refined his successional account and won a prize for it from the French Academy of Sciences in 1857. It appeared in print in 1858, in German, as *Untersuchungen über die Entwickelungs-Gesetze der organischen Welt* (Investigations into the developmental laws of the organic world; hereinafter, *Entwickelungs-Gesetze*),[10] the same year that Darwin, with Alfred Russel Wallace, gave the first public preview of his evolutionary approach.[11] The following year, Darwin's *The Origin of Species* appeared, which quickly and thoroughly overshadowed Bronn's work.

At the time of his initial contact with Bronn, Darwin had not yet read the prizewinning monograph, and he did not seem to be conversant with many details of Bronn's successional theory. He probably did not even read Bronn's review of the *Origin* very carefully, either, for even though he told Bronn that he had been quite pleased with it, he wrote to the geologist Charles Lyell that "The united intellect of my family has vainly tried to make it out.—I have never tried such confoundedly hard German: nor does it seem worth the labour."[12] This was a serious miscalculation, because Bronn would not only soon translate Darwin's book, but also append those notes of refutation and confirmation that Darwin had invited and that would steer the initial German debates over Darwinism into channels of Bronn's choosing.

Bronn's version of *The Origin of Species* appeared in 1860, mere months after the original. It was the first foreign-language edition on the market, and it immediately provoked debates and challenged German scholars to think about morphology, systematics, paleontology, embryology, and other biological disciplines in new ways. Such a process of transmission, translation, and historical change in science might once have seemed unremarkable, even to be expected, but historians no longer take the universality and easy transferability of scientific knowledge for granted. Linguistic, cultural, and geographic barriers now loom large, and science is seen as closely tied to the social and political interests of individuals, communities, and states, as well as to materials, assumptions, and practices that do not travel easily. The process of bringing Darwin's *Origin* to Germany and making it understood there clearly involved much more than a mere mechanical substitution of German words for English in a text.

I take for granted, however, that not all of science is local. Modern scientists do manage to understand one another reasonably well, because at least some of their technical vocabulary, assumptions, and practices are shared, and especially because they have experience with many of the same methods and objects of study. The nineteenth-century biologists to be discussed here—chief among them Darwin, Bronn, and Darwin's most famous German interpreter, Ernst Haeckel—all wrote great compendia of morphology and systematics, and all had an encyclopedic command of natural history, including experience with at least some of the phenomena most important in Darwinism, such as variation. Indeed, in some areas, Darwin and his German interpreters had more in common with each other than with present-day biologists and historians who try to read their work and translate it into modern terms. If the historical sources appear to differ, we must beware of jumping to the conclusion that the Germans mangled or misconstrued Darwin's obvious meaning. It may be that we are misunderstanding the historical sources because of the ways in which both Darwinism and its technical vocabulary in English and German have evolved. Existing historical accounts, which differentiate strongly between Haeckel's aberrant evolutionism and an Anglo-American mainstream, have often fallen into this trap by reading pre-Darwinian meanings into the German Darwinian's words. The extent of Bronn's understanding and support of Darwin's program, too, has been grossly underestimated, for similar reasons.

This book therefore reopens the question of how German Darwinism relates to Darwin's own version, and it examines the translation and early interpretation of Darwin's book and theory in detail. It focuses on the level of the key individuals involved in the process, while also considering the states of the various fields of pre-Darwinian biology in which they were trained and to which they wished to contribute, their aspirations for zoology, paleontology, and morphology as academic disciplines, their general scholarly ideals, and their standards of scientific evidence and explanation. It will show how these considerations, along with personalities and professional rivalries, played out in the interpretations, criticisms, and wordings of Darwin's theory.

The interpretive problems they had to solve ranged from the verbal (how could Darwinian concepts be expressed with the existing German technical vocabulary?) to the visual (what did all those British hunting dogs and fancy pigeons actually look like that were supposed to illustrate selection, common descent, and divergence?) and, of course, to the conceptual. Among the latter, one set of problems stands out as particularly troublesome and divisive and will be the leitmotif of this book. These problems have to do with striking the proper balance between the operation of fixed laws—which it was the job of the pre-Darwinian natural scientist to discover—and the possibility of unpredictable, historically contingent variation and creative change. How free were individuals or groups to vary all over the morphological map, or how constrained to stay within the limits of a taxonomic type, to climb a predetermined scale of nature, or to repeat ancestral patterns of development? Could evolution continually create truly novel forms, or did it only bring out preestablished potential? Was change driven and directed more by the internal constitution of the organism and the laws and mechanics of embryonic development, or by complex and unpredictable interactions with the external environment? Specialists in evolutionary biology continue to be vexed by comparable questions today, concerning the relative importance of developmental and phenotypic constraints, drift, epigenetic heredity, the origins of evolutionary novelties, and the relationship between development and evolution.

Unfortunately, a large body of secondary literature has placed the pre-Darwinian German morphologists squarely on the side of idealized archetypes, deterministic scales and laws of development, and limitations on variation and creativity. This "transcendentalist" or "developmentalist" bias is presumed to have carried over into the Darwinian period, giving rise to an aberrant evolutionism. It has been linked by various chains of association to the worst biology-based political ideologies of the twentieth century and it has kept the German school out of the overall picture of the history of Darwinism, from *The Origin of Species* through the modern evolutionary synthesis and beyond.

As I shall argue, however, the sources have been read in a one-sided manner, and the persistence of pre-Darwinian vocabulary has been mistaken for the persistence of pre-Darwinian interpretations. To Bronn and Haeckel, Darwinian concepts of variation and historical contingency were neither unanticipated nor unwelcome. The question for them was not whether to accept these unruly processes, but how much free rein to grant them, and how best to talk about them.

Bronn's Point of View

What Darwin missed in his initial reading of the confoundedly hard German of Bronn's book review was not any significant attempt at refutation. On the contrary, although Bronn expressed skepticism about the transformability of species, his tone was generally favorable. The problem was what Bronn was being favorable about. Upon reading the fine print, Darwin found Bronn's explanation of "natural selection" to be a far cry from what he thought he had said. Bronn had the struggle for existence pressuring organisms to "select" new environments and new ways of life, to which they subsequently became adapted, more or less as in Lamarck.

Darwin delicately corrected Bronn on this point, but over the course of the translation project, there were many pitfalls in the way of their mutual understanding, despite their parallel research interests. Not only was Darwin's English, as he said, "much condensed," and Bronn's German confoundedly hard, but the two naturalists had different goals, methods, assumptions, and standards of evidence and argument, particularly when it came to formulating and supporting historical inferences and laws of change. They had different, albeit overlapping, stores of empirical knowledge and experience to draw on. And finally, there were personal tensions between the two, at one and the same time allies in promoting a historical view of life, collaborators on a book project, and rival theorists. These tensions, too, colored Bronn's interpretations, often in contradictory ways.

The Darwin industry, when it has taken note of Bronn at all, has not taken these linguistic and interpretive pitfalls into consideration, let alone those that stand between nineteenth-century authors and twentieth- or twenty-first-century readers of either language. There is a tendency to snipe at Bronn for taking liberties with the text and misunderstanding Darwin's theory.[13] Bronn is presumed to have been so deeply mired in an outmoded German transcendentalism that he was incapable of making the shift to the new evolutionary paradigm and could not help but misrepresent it as a variation on familiar teleological models. Only a few authors have made the effort to understand Bronn's point of view,[14] but even they do not quite see him as a man of the 1850s and a reliable interpreter of Darwin.

One reason why Bronn has been underestimated is because he tried to convey Darwinian concepts in pre-Darwinian German terminology. Even though one can hardly expect him to use twentieth-century vocabulary and to make twentieth-century distinctions, historians have seized upon keywords such as *Vervollkommnung* (perfection, or progress toward it) or the language of developmental laws and forces to tie Bronn to older ideas about archetypes, embryonic models of change, and so on, even though he himself had argued against such things in earlier works. Haeckel's use of some of the same vocabulary has allowed him to be tarred with the same brush.

There were significant differences between Bronn's version of the *Origin* and Darwin's, to be sure, but Bronn's words must be interpreted in light of his other writings, in which words like *Vervollkommnung* or other morphological terms were already taking on new meanings in the 1840s, meanings that were further modified and specified in the Darwin translation. Haeckel, too, may seem to today's readers to have spoken an outmoded language of law, progress, and perfection, but he took over many keywords from Bronn, and was saying new things with them in the 1860s. Close readings of the German texts will show Bronn and Haeckel searching for—and more often than not finding—common ground with Darwin, rather than forcing him into an anachronistic framework of transcendental morphology.

Bronn's role in the history of evolutionary biology and the international reception of Darwinism has been further obscured by the fact that he would never quite be convinced of species transformation. Darwin himself did not give Bronn much credit as a contributor to evolutionary thought, as if he had never learned anything of value from a non-transformationist. He expressed incredulity at Bronn's claim to have anticipated aspects of his theory and to have published a "foreboding" of them,[15] and he passed over Bronn in his "Historical Sketch" of evolutionism.[16] This, I think, has prevented Bronn from being treated as a "forerunner" or early supporter of Darwin, despite their common interests in history, variation, and adaptation, and despite Bronn's evident importance for disseminating Darwin's theory and initiating discussion of it in Germany. Similarly, historians have tended to deny Haeckel his credit for being an early Darwinian. His account of evolution is considered unworthy of even being called "Darwinism." It is referred to as Lamarckism, "pseudo-Darwinism,"[17] or (even in English) *Darwinismus*,[18] and it is linked to linear developmental progressions, teleology, and unsavory political ideologies.

Here I will argue that Haeckel followed Bronn very closely on important matters of interpretation and terminology, and that my new reading of Bronn as a reformer and not a perpetuator of the old morphology necessitates a new reading and repositioning of Haeckel as well. As in Bronn's case, the continuities with transcendental morphology, linear scales of nature, and deterministic laws have been overstated. I will argue, in addition, that Haeckel not only took over Bronn's terminology and conceptions of historical progress, but also let Bronn's "notes of refutation" set his agenda for shoring up the case for Darwinian evolution. He crafted his own system of evolutionary morphology in such a way as to answer Bronn, in a sense thereby completing Bronn's translation and giving German Darwinism its origin.

The Squire and the Professor

The themes of this book, and my principal points of comparison between German and British biology and Darwinism, are best introduced by the following juxtaposition of Darwin and Bronn, their careers and intellectual commitments, which will be followed up in greater detail in chapters 2 and 3. The accounts of Darwin and Bronn will then set the stage for analyses of the Bronn translation in chapter 4, and Haeckel's use of it in chapter 5.

Consider first Mr. Charles Darwin, M.A. He was born in 1809 into a family of wealthy physicians and industrialists. He studied medicine for two years at Edinburgh and then went to Cambridge to prepare for a career as a clergyman. He made his scientific reputation as a field geologist and tropical explorer, with very little formal training in the sciences, but with a great deal of personal guidance from individual authorities. He never occupied a university chair or other professional position as a scholar or researcher, but was always a self-financed, amateur gentleman-naturalist. This was hardly unusual at the time, since English science was just beginning to become professionalized and gentlemen-naturalists still dominated the scientific societies and publishing market.

The gentlemen's network that helped Darwin get his start in biology was steeped in the tradition of British natural theology, with its intense focus on function, organization, and complex adaptations to both the physical environment and to communities of interacting species. Although Darwin was also exposed to French and German ideas about species transformation and special formative laws and forces (principally at Edinburgh, but also through Richard Owen in the 1840s), his 1859 argument followed the line of his Cambridge mentors by rejecting such devices as explanations of complex adaptation and especially coadaptation. Neither chance nor blindly deterministic laws, he said, could ever explain such flexible and context-sensitive phenomena.[19] Either the free creative activity of a personal Designer had to be invoked, or else some new mechanism had to be discovered.

After his five years of travel aboard H.M.S. *Beagle*, Darwin settled down to the life of a gentleman on a country estate, and as he developed his theory of natural selection, he supplemented his knowledge of natural history with perspectives from the landed gentry: the latest methods of scientific agriculture and practical breeding, especially of hunting dogs and fancy varieties of pigeons. Also important were influences from the world of the entrepreneurial British middle class, such as developments in technology and industry, and ideals of progress through competition and inventiveness. In making his argument for natural selection, Darwin argued by analogy to familiar forms of human artifice and creativity. Natural selection was like an enterprising breeder, creating order and purpose out of nature's random variations.

Darwin has been celebrated for providing precisely the alternative to chance, law, and design that was unavailable to Paley, and making it possible to be "an intellectually fulfilled atheist."[20] Darwinian natural selection was neither purely random nor strictly lawlike and deterministic. It combined chancy variations with predictable Malthusian overproduction, struggle, and heredity into a complex that on the whole was neither a matter of chance nor necessity, yet was capable of explaining the same phenomena as Paley's Designer, and more besides.

But even before Darwin, other ways had been explored of explaining life's complexities naturalistically,[21] and the problem of how best to do so was especially acute for Bronn and the earlier German naturalists to be analyzed in chapter 1. These naturalists included some of the first professional "biologists," mostly state-supported researchers in a modernizing university system, and they were deeply concerned with the status of their field among all the other university disciplines. They continually discussed what was required to make the study of life into a proper intellectual pursuit, or *Wissenschaft*, distinct from both descriptive natural history and theology. To that end, they moved away from direct references to divine intervention, and away from eighteenth-century theories of "preformation" that might imply that present-day biological structures had been planned and prepared at the Creation.

Instead, the most promising naturalistic approaches tried to borrow conceptions of law and force from physics and apply them to biological problems. The laws had to be given some flexibility, some capacity for responding to the environment or to conditions within the organism, and they had to be made to interact with other laws and forces to produce complex and varied results. Darwin's theory had to prove itself clearly superior to such systems of biological laws and forces. It was not embraced unreservedly until the causes of variation could be clarified and the right balance struck between chance and necessity.

Darwin's German counterpart was the Herr Hofrat Professor Dr. Heinrich Georg Bronn. As the very name proclaims, he was no country gentleman, but a professional scholar and high-ranking civil servant. He was born in 1800 in a small town near Heidelberg, where his father was a forestry official. He was the fifth of seven children and was raised a Catholic. He studied natural history and *Kameralia* (cameral studies, the track for civil servants) at the University of Heidelberg, and he earned his *Habilitation* (*venia legendi* or postdoctoral qualification for university teaching) there in 1821, in the fields of applied natural history and *Enzyklopädie der Staatswissenschaften* (encyclopedia of the political sciences). Remaining at Heidelberg, he taught courses on forestry, agriculture, technology, applied natural history—apparently with an emphasis on mineralogy, mining, and soils—as well as topics in zoology and paleontology.[22]

Bronn was therefore not a typical academic zoologist. He got his training not in anatomy or medicine, as was more usual, but in natural history and cameral studies. The cameralist's perspective on the natural world was probably best expressed in Bronn's discussions of Malthus, population growth, and resource limitation; of practical breeding as a source of knowledge about environmental effects and variation; and of forest soils and other environments that both shape and are shaped by living communities. Bronn also sometimes seemed to bristle at the conventional zoological wisdom that subordinated paleontology to better-developed but ahistorical fields of research like morphology and embryology. He made a point of grounding his laws of historical change in his fossil data, denying their derivability from the study of living organisms alone.

Bronn rose through the ranks at Heidelberg, becoming extraordinary (*außerordentlicher*) professor for *Staatswirtschaft* (public finance) and natural history in 1828, and taking over the Zoological Cabinet and the primary teaching duties in zoology in 1833. In 1837, he was promoted to full professor or professor-in-ordinary (*Ordinarius*) in both zoology and applied natural history. He became Germany's most distinguished paleontologist, known for fieldwork in Italy and throughout Western Europe, identifying and sequencing strata of sedimentary rock and the fossils they contained. He published numerous works on zoology, pale-ontology, and pure and applied geology. He edited or coedited the *Neues Jahrbuch für Mineralogie, Geognosie, Geologie und Petrefaktenkunde* (New yearbook for mineralogy, geognosy, geology, and fossil studies)—in which he reviewed *The Origin of Species*—for over thirty years, and he launched the well-known series of taxonomic reference works, *Classen und Ordnungen des Thierreiches* (Classes and orders of the animal king-dom) which has been continued long after his death.

Bronn's major work on organic history, *Geschichte der Natur*, published in three thick volumes in the 1840s,[23] balanced a detailed study of the fossil record with a careful critique of its incompleteness and its biases. It pro-vided a wide-ranging, quantitative account of biogeography, including the small amount of data that was then available on fossil species distribu-tions outside of Europe. It analyzed the methods and accomplishments of plant- and animal breeders, using some of the same English sources avail-able to Darwin, and asking some of the same questions about the causes of variation and the production of varieties. It compared domestic with natural variation as well, and in so doing, it undermined some conven-tional assumptions about the fixity of taxonomic types, especially of the higher taxa, over geological time. It expounded upon the universality of Malthusian overproduction and the enormous loss of life that was the inevitable price for maintaining the balance of nature, and it even described existence in nature as a struggle, in which the maladapted would go under. It discussed the ubiquity of adaptation to the environment, including phys-ical and biotic, local and regional conditions, and the complex, reciprocal effects of species on species, and species on the landscape. Bronn showed an appreciation for the appearance of design in nature and treated natu-ral theology with respect, even as he moved, over the course of his career, toward a naturalistic alternative to the designing Creator.

In contrast to Darwin, who had natural selection impose order and apparent purpose on nature's random variations, Bronn explained how nature could be orderly and law-governed yet still produce a wild variety of landforms, mineral forms, plants, and animals. He saw variation as the product of nature's laws, not as raw material to be further sorted and shaped. Bronn's was a successional theory, not an evolutionary one. Instead of connecting the dots and drawing family trees, Bronn left fossil species disconnected and unrelated by heredity. Old species were driven to extinction as environmental changes gradually made their adaptations obsolete, and unidentified creative forces or processes caused new species to come into being that were both better adapted and morphologically more advanced. Bronn noted that new species often resembled the ones they succeeded, but that was because they were formed by the same laws and forces, under the influence of similar—but of course not identical—local conditions.

In one sense, however, Bronn's theory was more thoroughly evolutionary than Darwin's: it had the Earth evolving, too, gradually and progressively, in contrast to the steady-state model that Darwin took over from Charles Lyell. Ever since its formation as a uniform, molten sphere, Bronn argued, the Earth had been cooling. It had crusted over, contracted, cracked, and developed increasingly diverse and hospitable environments on its surface. Life had to change constantly in order to keep up with a dynamic Earth, and Bronn criticized purely biological evolutionists like Lamarck for treating the Earth as static and missing the very reason why organic change was necessary.

In Bronn's system, each species lived until gradual geological and environmental change, including changes in the flora and fauna with which it interacted, made its survival impossible. There were no catastrophic mass extinctions and subsequent mass creations, as older theories had it, for example that of Louis Agassiz, whom Bronn was most keen to refute. Bronn's detailed stratigraphic data consistently showed that species came into existence and went extinct individually, rather than in synchrony, and as they came and went, there was a continual, gradual, and progressive turnover in the composition of the world's flora and fauna. Life became more advanced by several measures (also devised by Bronn), and by Bronn's logic, such organic progress followed *necessarily* from progressive change in the Earth and the principle that species must always be well adapted to the conditions they would actually encounter. Bronn argued that necessary, quantifiable laws were behind the observed patterns of organic change, and the goal of his research was to discern those laws, even if it could not identify the mechanism by which new species originated.

Problems of Translation

Bronn and Darwin never met, but they communicated—and miscommunicated—in writing, including their published scientific works and letters exchanged between 1859 and 1862. Given their contrasting backgrounds, positions, and conceptions of biology and science, it is easy to see that their mutual understanding might fail, not only because of linguistic problems. Recent literature on translation recognizes that the process involves more than the mechanical substitution of words in a text, and sees translators and interpreters as authors in their own right.[24] This approach opens new possibilities for interdisciplinary bridge-building between the history of science and literary or cultural studies, and it can enrich the historiography of science as well. It encourages a view of scientific theories not as fixed propositions to be accepted or rejected as originally stated, but as historical entities that change through time and across national boundaries. In the case of Darwinism in particular, as David Hull has argued, it is impossible to find any unchanging essence of the theory. There were significant differences of interpretation even within Darwin's original circle of British supporters.[25]

The analysis of translation is also an important complement to a growing literature connecting scientific knowledge to particular places,[26] movable objects, including, but not limited to, books[27] and national traditions or "styles."[28] These approaches call new attention to the process of transmission in science and assign greater independence to the recipients of foreign ideas and influences. In the case at hand, we will see that Darwin's German interpreters, to some extent, made his theory their own and turned it to their own purposes. But we must also beware of exaggerating the independence of the translation or interpretation from the original. Much may have been lost or changed in translation, but much was also communicated successfully. Some of the deviations of Bronn's translation and later interpretations by Haeckel from Anglo-American understandings of Darwin are more apparent than real. In any case, they will be treated here as matters of historical interest, rather than worthless errors and distortions.

The most obvious translation problems that Bronn had to deal with were the linguistic ones: what were the best choices of German words for Darwin's English terms, especially when these were being used in novel ways? What did Darwin mean, for example, when he said that species "progressed" or became "improved" or "perfected" by natural selection? By using these terms, was he alluding to, and perhaps endorsing, pre-Darwinian ideas about forms on a linear scale of nature? Or was he appropriating existing language for new purposes? Did the German "*Vervollkommnung*" have the same range of possible meanings, and could it be extended to cover Darwin's new ones? What about "the preservation of *favoured* races in the struggle for life" in Darwin's subtitle? Who favored them and why? Could it have been because they were more *vervollkommnet*?

The old morphologists did not have a monopoly on that word, nor did they all use it in the same way. Bronn put his own spin on it in the 1840s, when he analyzed *Vervollkommnung* into components such as increasing specialization and differentiation of organs or parts, concentration of functions into fewer serially repeating segments or organs, and internalization and protection of gills, sense organs, and other important surface features. In the 1850s, he began to relate these measures of progress and perfection to other authors' conceptions of "physiological division of labor." Most, if not all, of Bronn's measures had obvious potential for improving functionality and survival, providing Bronn with a way to relate *Vervollkommnung* to adaptation and Darwinian competitive improvement and to justify using the German word. Darwin himself made liberal use of "progress" and "perfection" to mean what he wanted them to mean, usually improved ability to compete in the struggle for life, but sometimes also specialization and division of labor, which he, too, linked to competitive improvement. In the context of the German *Origin*, then, did the old morphological language really change Darwin's meaning, or did Darwin and Bronn together change the language? My answer is that Bronn and Darwin worked together to redefine progress, perfection, and *Vervollkommnung*, and that it was quite reasonable for Bronn to use the old terminology. What neutral language was there for him to use in its place? Or what better one for communicating Darwinian ideas to German morphologists?

Some of Bronn's translation problems were not strictly linguistic, but reflected the authors' differing experiences with nature, Bronn's more paleontological and European-centered, and Darwin's more global and oriented toward living organisms. Bronn's knowledge was encyclopedic, but still he did not know exactly the same animals and plants with the same degree of familiarity and by the same names as Darwin. The problems were most acute in discussions of artificial selection, where neither names nor breeds were standardized internationally. What varieties was Darwin actually talking about? Even when Bronn could find the right German names for them, had he or his readers ever seen them, and could they fully appreciate the point that they were supposed to illustrate?

Still other translation problems had to do with Bronn's and Darwin's contrasting conceptions of biology as a scientific discipline, and their differing ideas about what any successful theory of species transformation would have needed to explain, and what would have constituted a proper scientific explanation. Bronn might have traveled less widely than Darwin, but he dug more deeply into the past and spent more time trying to reconstruct the histories of individual species and higher groups, faunal and floral assemblages, and the environmental conditions they encountered. He expected the same efforts at historical reconstruction from Darwin, but the *Origin* avoided specific claims about any species' ancestry. Emphasis was on the mechanics of evolutionary change, rather than its actual course.[29] In contrast, Bronn did without a mechanism of change and sought patterns and laws instead, and he sought them in his large catalog of fossil data, shunning hypothetical histories. In the German *Origin*, Bronn continually tried to depict natural selection as a law or force, and he complained about Darwin's use of language that seemed to personify natural selection and give it decision-making capabilities.

Further problems were of a personal nature. There are places in the German edition where Bronn seems to have reacted to perceived slights in Darwin's text, for example where a relevant work of his went uncited or, even worse, where Darwin cited British authorities for points Bronn had made, or argued with British opponents over issues that Bronn thought he had clarified. A few of these provocations elicited brief objections in footnotes to the main text, but other offending passages became targets for extended criticism in Bronn's appended commentary. Darwin's preoccupation with his network of British naturalists proved detrimental to his relationship with Bronn, whom he did not treat with the same deference in the text.

The Legacy of Idealistic Morphology

The breadth of Bronn's common ground with Darwin, his interests in zoology, geology, forestry, agriculture, and applied science, his concerns with adaptation and environment, and his rejection of embryological analogies to organic history do not fit easily into the picture of pre-Darwinian German biology that was bequeathed to us by E. S. Russell and other pioneering historians of biology, and reinforced by later accounts of German Darwinism. This picture centers on morphology and the predominance of transcendental or idealistic interpretations of form. It also links the success of Darwinism to its compatibility with the older, idealistic interpretations.

According to Russell, German transcendentalism, *Naturphilosophie*, and related approaches associated with Étienne Geoffroy St. Hilaire in France viewed plant- and animal forms as expressions of ideas, plans, or archetypes that had no material existence, but occupied a separate Platonic realm or perhaps the mind of God. Such transcendental types were arranged on an idealized, linear scale of nature in order of increasing complexity, and that scale became the template for all morphological sequences. Taxonomic groups were arranged in parallel to the scale of nature. The real changes observed in the developing embryo were said to recapitulate the sequence of lower adult forms. Similarly, species transformation or succession in the fossil record was understood to be analogous to embryonic development and was also presumed to run in parallel to the scale of nature. Adults of past or lower species were equated with embryonic stages of present-day ones, a view that Russell called the Meckel-Serres law of development, after the comparative embryologists Johann Friedrich Meckel in Germany and Étienne Serres in France.[30]

Russell contrasted this view of embryology, evolution, and the animal scale as predetermined, linear sequences of increasingly complex adult forms with the supposedly more modern account by Karl Ernst von Baer. Although still a transcendentalist, in the sense of positing nonmaterial ideas or essences that guided development toward a goal, Baer was a good transcendentalist, in Russell's book, because he rejected the linear animal scale in favor of a branching one, and because he redefined the direction of development from ascent along the scale to differentiation within a branch. The process of development now took the embryo from the general to the specific, from the more-or-less homogeneous egg to a gradually more recognizable member of its class, order, family, and ultimately species. Along the way, it never had to pass through the adult forms of any other species, and no scales of nature or historical legacies determined its evolutionary future.

According to Russell, Darwinism was easily assimilated into morphological thought because of its strong intellectual affinities with the wrong kind of transcendentalism, with Meckel's more than Baer's. The process of assimilation was initiated by Darwin himself, who summarily and superficially redefined the terms of the transcendental morphologists, in just one chapter of the *Origin*, without really trying to reform the field: "The current morphology, Darwin found, could be taken over, lock, stock, and barrel, to the evolutionary camp," Russell wrote. Archetypes were not done away with, or even brought all the way down to Earth from the transcendental realm; they were merely relabeled as hypothetical common

ancestors. They remained just as fanciful and irreal as they had been under transcendentalism, and they continued to play the same role in accounting for within-group similarities, guiding development, and determining form. The scale of nature became the evolutionary lineage, and the idea of embryonic parallelism or recapitulation was revived as well, mostly by Haeckel. Russell's thesis, in short, is "that the coming of evolution made surprisingly little difference to morphology, that the same methods were consciously or unconsciously followed, the same mental attitudes taken up, after as before the publication of the *Origin of Species*."[31]

Russell's thesis about the continuity between German idealistic morphology and Darwinism has been very influential, especially as applied to evolutionary morphology in Germany. This view has several problems, however. One is that it begins with an inaccurate account of pre-Darwinian morphology. As I shall argue in chapter 1, the linearity of the scale of nature and the parallelism of embryos with lower adults were not as firmly entrenched as Russell would have it, and early morphologists such as Meckel did not draw the morphological parallels that he attributes to them. Meckel did not have embryos repeat the adult forms of lower species in such a rigid linear sequence, but described a process of differentiation comparable to Baer's. Russell's dichotomy between the Meckel-Serres progression and Baerian differentiation is a false one, based on Baer's own overinflated claims of originality; and Russell's association of recapitulationism, progress, linearity, and transcendentalism is misleading.

A further problem with Russell's account is that there was much more to pre-Darwinian German biology than morphology, and much more in the *Origin* than one chapter on that subject. An analysis of the German Darwin reception must consider other subject areas and approaches, and must also recognize that even within "transcendental morphology" there were many different conceptions of types, scales of nature, models of development and evolution, and other explanatory devices. Finally, Russell's story skips over the 1840s and 1850s, when transcendentalism was in decline, and it fails to distinguish late pre-Darwinian directions from the earlier ones. The story serves Bronn as poorly as Darwin and Haeckel by lumping them all with the transcendentalists of the 1810s and 1820s.

A Revisionist View of Haeckel

Under the influence of Russell, the work of Haeckel and his sometime collaborator Carl Gegenbaur toward a reform of morphology has been seen as a half-hearted effort to save the old transcendental concepts and

continue to focus on types rather than on Darwinian variation and adaptation.[32] They injected just enough Darwinism into the subject to make archetypes into ancestors, but continued to construct them unrealistically and to force evolution onto a linear track. Haeckel's famous "biogenetic law" of recapitulation is treated as a revival of Meckel's supposed parallelism of embryos with a linear scale of lower adults.

This view is difficult to reconcile with Haeckel's convert's zeal for Darwinism, his stated ambitions for unifying biology within an evolutionary framework, and especially his stance against religious dogma, which required evolutionary contingency and continual creativity in nature. If he was such a determinist or idealist at heart, he must have been extremely insincere or self-deluding, and he must have kept up the charade for over half a century. As I shall argue instead in chapter 5, Haeckel struck a balance between the creative and conservative processes in evolution, the former leading to adaptation, diversity, and progress, and the latter maintaining family resemblances and preserving historical sequences of forms in the embryo.

Haeckel saw in Darwin's dual mechanism of variation and natural selection the only hope of providing fully naturalistic explanations of adaptation, morphology, and taxonomy, and purging the sciences of religious dogma and mysterious transcendental ideas and purposes. In tandem with Gegenbaur, he initiated a research program of reconstructing evolutionary history from comparative studies of anatomy and embryonic development. This was his answer to a criticism of Bronn's concerning Darwin's inability to document any particular line of descent. Haeckel's recapitulation theory provided a method and a justification for identifying remnants of ancestral forms and their successive modifications in the embryo, without invoking archetypes or idealized scales of nature. Haeckel envisioned a unified system of the biological sciences in which every subdiscipline would contribute to the project of reconstructing the history of life and, in return, derive the benefits of historical explanation. Indeed, his program of unification and evolutionary explanation extended beyond the traditional realm of the life sciences to encompass the study of the human mind, language, aesthetics, and aspects of philosophy and religion as well, on the grounds that, if Man was a product of evolution, then so were his mental and cultural features.

Haeckel's Darwinism was therefore more than just window dressing for the older morphology, and indeed more than just a biological theory. It was also part of his philosophy of "monism," a unified worldview that reduced mind and spirit to natural properties of matter, treated human

affairs as part of the organic realm, opposed transcendental religion (particularly Catholicism), and preached a biological basis for ethics. It attracted a bewildering spectrum of political activists and social thinkers, from future fascists and National Socialists to liberals, Marxists, advocates of women's rights and sexual reform, atheists, Freemasons, and assimilated or converted Jews, all in search of a scientific basis for social thought and action.

The feminist Helene Stöcker found Haeckel's viewpoint attractive for providing a secular, scientific approach to ethics that denied original sin and could counter conservative Christian objections to women's emancipation. For the sexual reformer Magnus Hirschfeld, Haeckel's authority brought human love and sex under the purview of science and made homosexuality a biological condition instead of a form of moral depravity. Socialists, too, could make use of Haeckel's anticlericalism, materialism, and beliefs in progress and in the inheritance of acquired characteristics.[33]

Specifically, Haeckelian monism held that the world was made up of a single fundamental substance, "matter," a view that Haeckel contrasted with the various forms of "dualism" that recognized separate, nonmaterial entities such as mind, soul, spirit, vital forces, and deities. This was not at all new or surprising to his German audience, who still remembered the materialism controversy of the early 1850s, sparked by Carl Vogt, Ludwig Büchner, and Jacob Moleschott. Vogt's pronouncement that thought-production by the brain was no different in principle from urine-production by the kidney was a subject of public discussion that had polarized the German Society of Naturalists and Physicians in 1854 as it debated whether the soul had any place in physiology and psychology.[34] However, Haeckel's monism differed from the materialism of Vogt, Büchner, and Moleschott in its conception of "material." The monist's material was at once matter, mind, and energy, for it was the one and only substance. Its rudimentary mental qualities helped explain the specificity of chemical attraction and repulsion, as well as memory and heredity. This solution allowed Haeckel to blur the distinctions between the animate and inanimate, the mind and the brain, and to make life—including human life—just another natural phenomenon.[35]

Evolution was just another natural process, driven by changes in the physical environment, with no more plan or purpose than the weather or anything else in the inanimate world.[36] When explaining the role of the environment in evolution, Haeckel assumed that the environment could modify an organism, either directly or by affecting its behavior and letting the use or disuse of an organ modify its form. The effects were presumed to be heritable, at least under some circumstances, a principle known as the inheritance of acquired characteristics. Such environmental effects had been at the core of Lamarck's theory of evolution and have led some historians and scientists to dismiss Haeckel as a Lamarckian. That may be justifiable; but Darwin's original theory also made use of such effects, and it does not seem reasonable to make the one into a Lamarckian and not the other. As will be argued in chapter 5, Haeckel applied such environmental effects in a Darwinian fashion by making them a source of variation. To be sure, it was not "random" variation in the modern sense, as it was rather biased toward producing adaptive and progressive changes. Still, these variations were the raw material upon which natural selection had to act, and this added selective step made the resulting process more like Darwin's than Lamarck's.

In prosecuting his case against the clerics, and simultaneously answering one of Bronn's criticisms, Haeckel removed every vestige of a Creation from his Darwinism. Species were formed neither directly by divine intervention nor indirectly by predictable, teleological processes set into motion long ago by the Creator. Evolutionary changes had to be triggered by unpredictable forces and changes in the physical environment. There was no unbroken chain of purely biological causes that accounted for evolution and could be traced back to the Beginning.

But, again, that was only one side of Haeckel's account. Haeckel had to strike a delicately calibrated balance between the contingency and unfettered creativity of evolution on the one hand, which he needed against the clerics, and the historical continuities and constraints that made the developing embryo repeat at least some of its evolutionary past, which enabled Haeckel to reconstruct phylogenies. He was most vigilant in defending this balance against laws that directed variation and evolution, but he also upheld it against models of heredity or development that, in his opinion, overemphasized causes of change within the embryo and neglected the external environment. Consequently, Haeckel rejected most forms of orthogenesis, or directed evolution, but also the mechanistic embryology of Wilhelm His, August Weismann's germplasm theory, the *Entwicklungsmechanik* or developmental mechanics of Wilhelm Roux, and even saltational theories and early conceptions of mutation.

Explaining the Haeckel Reception

At least two generations of biologists were inspired by Haeckel's writings to enter the field and to apply evolutionary concepts in various subdisciplines. The many contributors to the two-volume tribute to Haeckel, issued for his eightieth birthday in 1914, testify to the inspiration he gave for the use of Darwinian ideas not only in biology, but also in politics and social-reform movements of every description.[37] Even younger biologists, for whom Haeckel was no longer such an icon, continued to encounter his ideas and intellectual presence until well into the twentieth century. As the geneticist Richard Goldschmidt acknowledged, somewhat grudgingly, "The present generation cannot imagine the role he played in his time, far beyond his actual scientific performance."[38] Unfortunately, even before his death in 1919, historians and biologists began to lose sight of Haeckel's scientific performance altogether and to hold it in even lower esteem than Goldschmidt.

Haeckel's reputation has also suffered, in part, because the definition of Darwinism has changed to exclude the inheritance of acquired characteristics. Twentieth-century evolutionists, whose main concern was to synthesize Darwinism with population genetics, therefore labeled Haeckel a Lamarckian and distanced themselves from him. They had little interest in his positive contributions to the foundations of phylogeny, evolutionary morphology, and a phylogenetic taxonomy. But Haeckel's worst misfortune was that the first and most widely read twentieth-century histories of biology were written by biologists with anti-Darwinian axes to grind, who were also anxious to rid morphology of what they thought was Haeckel's pernicious influence.

From very different points of view, the Swedish morphologist Erik Nordenskiöld and the British fisheries biologist Russell both rejected the theory of natural selection, the use of evidence from comparative morphology to establish evolutionary relationships between species, and Haeckel's monism. Their accounts of Haeckel were vague, dismissive, and contradictory, but they connected him strongly to *Naturphilosophie* and transcendental morphology, which set the tone for the reception of Haeckel's work by later historians.

According to Nordenskiöld's *History of Biology*, Darwinism was not only wrong, but also already completely rejected by the 1920s, and Haeckel, as the leading proponent of a defunct theory, was little more than a historical curiosity. The proper goal of morphology, Nordenskiöld said, was to explain organic forms as products of the laws of nature and not as historical accidents. If the laws produced forms that turned out to be adaptive or purposeful, that did not require a separate explanation. Organisms simply were as they had to be. Hence, Haeckel's enthusiasm for natural selection as the only possible naturalistic explanation of adaptation was misplaced. It was the answer to a superfluous question. Further, any resemblance between species indicated only that they were formed according to the same laws; it could not be taken as evidence for their common ancestry. Haeckel's project of reconstructing phylogenetic trees based on morphological resemblances was, therefore, completely unscientific. Nordenskiöld did not see much difference between Haeckel's quest for hypothetical ancestors and the early-nineteenth-century quest for transcendental ideas and types, or between monism and any other speculative system building, so he had no trouble labeling Haeckel a German idealist.[39]

Russell's 1916 assessment was equally negative and also contradictory, with perhaps a touch of wartime resentment against anything German. He ascribed to Haeckel both intransigent materialism and transcendental idealism; unjustified opposition to teleology and thinly disguised developmentalism; and a penchant for absolute distinctions along with an excessive amount of hedging. Russell himself was an unabashed vitalist, who thought morphological change was directed by the will and purpose of the individual organism, and from that perspective he labeled Haeckel a materialist and a determinist. On the other hand, it made little difference to Russell whether one rejected biological purpose altogether, like Haeckel, or removed it from the organism to some divine or Platonic realm, like the transcendentalists. Russell equated ideal types with hypothetical common ancestors, and rejected both because they reined in the creative power of life in order to maintain either typological or family resemblances. Hence Russell was also able to tar Haeckel with the same brush as his transcendentalist opponents and label him an idealist as well as a materialist.[40]

The main point of contention among Haeckel, Russell, Nordenskiöld, and later critics is the relative importance of the rule of law and historical contingency in embryology and evolution. Does a long chain of deterministic, mechanical causes link every individual to its ancestors and tend to constrain the embryo to follow established pathways of development? Nordenskiöld, following Haeckel's old opponent Wilhelm His and experimental embryologists of the *Entwicklungsmechanik* school, would have said that such a causal chain existed and that Haeckel tried unjustifiably to deny it.

But Haeckel required this kind of historical constraint on embryological development, too, in order to justify his methods of reconstructing evolutionary history: the effects of the phylogenetic past had to be detectable in the present embryo. He got no recognition for this from Nordenskiöld, but he fell afoul of Russell, who emphasized the power of life to adapt form to function, and its freedom to evolve as its needs dictated, without being constrained to recapitulate the past. Russell attributed to Haeckel a "Prussian mania for organization, for absolute distinctions, for iron-bound formalism," and he complained further that "A treatment less adequate to the variety, fluidity and changeableness of living things could hardly be imagined."[41]

Yet, Haeckel also required that unpredictable environmental influences and other historical contingencies intervene continually and let the embryo break free from its past in order to vary and adapt in novel ways. It served his ideological purposes and polemics against religion to have nature create forms that one could not have predicted from knowledge of the Creation and the laws of development and heredity. Few authors have noted the indeterminism and open-endedness of Haeckel's system and its contrast with the developmental laws of the older school,[42] a crucial point for an understanding of Haeckel's role in transforming German biology and the theoretical basis for his embryology, comparative morphology, and taxonomy.

This open-endedness and indeterminism earned Haeckel no credit from Russell, but it fell afoul of Nordenskiöld. Nordenskiöld argued that scientists should be looking for predictable, mechanical causes of embryonic change, rather than random events in the evolutionary past. He wrote: "In reality, this mechanical . . . side of embryonic development is of great importance, though Haeckel quite overlooked the fact in his anxiety to explain natural creation."[43] Instead of the rigid Prussian of Russell's book, Nordenskiöld's Haeckel was an undisciplined, romantic dreamer, prone to mystical speculations.

Most later authors follow one or both of these lines of interpretation—but more often Russell's—to place Haeckel in the idealist camp and deny him his claim to being a Darwinian. On Nordenskiöld's authority, Daniel Gasman described Haeckel's biological works as German Romantic *Naturphilosophie*, and denounced *Naturphilosophie* as charlatanism and quackery. From this, he proceeded very simplistically to derive all that was bad in later German intellectual and cultural history—social Darwinism, eugenics, pseudo-scientific theories of race, and ultimately Nazi ideology—from Haeckel's influence.[44]

Stephen J. Gould's *Ontogeny and Phylogeny* endorsed Gasman's assessment of Haeckel's irrational mysticism as a contribution to Nazism, but mainly took Russell's viewpoint as it characterized Haeckel as a "biological determinist" and an idealistic morphologist. Gould reinforced Russell's identification of Haeckel's recapitulationism with Meckel's, and

Meckel's with a linear sequence of lower adults. By Gould's reasoning, Haeckel's recapitulation theory required every individual to go through an inflexible sequence of developmental stages, determined by ancestry. Variations could be added on only in a limited way at the end of development, a process Gould called "terminal addition." Gould traced various forms of biologically based discrimination, including National Socialism, to this sort of deterministic thinking.[45] Actually, as will be shown in chapter 5, Haeckel had no such term as "terminal addition" and no such restriction on the timing of developmental variations.

Building on Russell, Gasman, and Gould, Peter J. Bowler further embellished Haeckel's purported idealism and determinism and created the category of "pseudo-Darwinism" for it. Bowler's main reason for demoting Haeckel's Darwinism was Haeckel's use of an analogy between evolution and embryonic development. It was not the same analogy as Meckel's, but to Bowler it nevertheless recalled the old idealists or "developmentalists," and Bowler therefore foisted their supposedly linear, deterministic, and teleological conception of development on Haeckel and extended it to Haeckel's picture of evolutionary change. Bowler argued that Haeckel's recapitulationism "encouraged the belief that evolution shares the progressive and teleological character of individual growth."[46] Bowler also endorsed Gasman's derivation of National Socialism from Haeckel's evolutionary thought, so he kept German Darwinism on a separate line of intellectual development from the more correct Anglo-American variety, which it managed temporarily to "eclipse" in importance, until the modern evolutionary synthesis of the 1930s and 1940s laid it to rest.

Consciously or not, these authors have read Haeckel under the assumptions of the *Sonderweg* theory, the idea that German political and social history followed a pathological "special path" to modernity, which led to National Socialism instead of British- or French-style parliamentary democracy. Typically on this view the errors and failures of Germany's middle-class liberals are blamed for setting the deviant course. Applied to biology, *Sonderweg* logic has traced the deviant path from the "biological determinism" of Haeckel and the earlier idealists to Nazi racial ideology. The original *Sonderweg* theory or theories have been under fire from social historians since the 1980s for arbitrarily taking French or British history as the norm from which Germany deviated, and for focusing on what German liberals failed to do instead of what they actually did.[47] But the parallel story of German Darwinism, which has many of the same shortcomings—particularly the teleology and the anachronistic mining of nineteenth-century sources for proto-Nazi ideas—remains to be reexamined.

Newer Historiographic Directions

Important efforts have already been made to study German morphologists and evolutionists on their own terms, taking them off the *Sonderweg* or otherwise reconnecting them to the international mainstream. Timothy Lenoir made a groundbreaking early move toward differentiating among German approaches to biology, and he found more than just idealistic morphology and Romantic *Naturphilosophie*. He filled out and corrected a good deal of Russell's caricature with an analysis of what he called "teleomechanism" and of morphologists' uses of physicalistic laws and forces.[48] Lynn Nyhart's cross-sectional and cross-generational survey of German morphology as a discipline provided a long-needed alternative to Russell's account. Although it retained his periodization into transcendental, evolutionary, and experimental morphology, it brought out a greater diversity of approaches and institutional contexts within each period, and it showed that the lines could not be drawn quite as clearly as Russell had them. Most important, for my purposes, Nyhart challenged Russell's (and Gould's and Bowler's) too-easy transition from transcendental to Darwinian interpretations of form.[49] She showed that German morphologists were already searching for alternatives to the old idealism before Darwin.

Nyhart described the changes in German biology as an intellectual and institutional "splintering" that began in the 1840s, when research specialties and university chairs in zoology began to multiply. Cell theory, paleontology, biogeography, physicalistic trends in physiology, and functional concerns in morphology all challenged the very conceptions of biological science with which the older morphologists had worked.[50] These newer developments, including Bronn's ideas about interpreting the history of life, were important elements in the intellectual context for the German reception of Darwinism.

In contrast, Robert J. Richards has been building on Russell's questionable foundation and reasserting the intellectual ties among Haeckel, Darwin, and transcendental morphology and embryology. Unlike Russell, however, he has used the association not to discredit Darwin and Haeckel, but to rehabilitate the older morphological camp by having Darwin emerge from it. Richards also differs from Russell by arguing that Darwin did not merely subsume the concepts of transcendental morphology into his new system in a superficial way, making hypothetical archetypes into hypothetical ancestors and the scale of nature into the evolutionary lineage. He has made Darwinian evolutionism a much more direct outgrowth of

German idealistic morphology. According to Richards, the recapitulation-ism of the idealists provided the main conceptual or metaphorical link between embryonic development and species transformation. Once reca-pitulationists went so far as to draw their parallels from the embryo to the scale of nature, it was but a small step from there to new parallels with the sequence of forms in the fossil record and actual ancestry.[51]

The steps from the embryo to the ideal scale to the historical lineage might indeed look small today, when evolutionary interpretations are commonplace in biology, but what evidence is there that pre-Darwinian transformationists actually took those steps with such ease, and in the sequence suggested? As I shall argue in chapter 1, some recapitulationists did indeed allow for species transformation, but they did not come to this idea by way of the small step that Richards envisions. They did not cite embryological or comparative morphological evidence in support of species transformation. They did not think that morphological similari-ties necessarily resulted from genealogical relationships, but allowed for multiple historical paths to the same form.

A further problem that Richards has inherited from Russell is that through all its vicissitudes, he treats the archetype as *the* archetype, no matter where it was, what it consisted of, or how it acted, even after it was reconceived as an ancestral form. Such blurring of distinctions makes it easy for Richards to explain the generally positive German reception of Darwinism, because, as an extension of older German ideas, the theory would never have seemed new or alien in the first place.[52] But it should be clear that the historians are the ones who make this connection. The sci-entists whose work is being lumped together made distinctions among the various possible usages of "archetype," "ancestor," "*Vervollkommnung*," and developmental analogies, and we ought not to sweep those histori-cal distinctions aside. Attention to these distinctions will show German morphologists—most notably, Bronn—moving away from transcendental interpretations. Darwin's positive reception is better accounted for by his ability to solve the problems of the 1840s and 1850s than by any per-ceived affinity to an older school. In fact, such an affinity would have been rather to his disadvantage by 1859.

The latest and most comprehensive analysis of Haeckel's work is by Mario Di Gregorio, who seems to be seeking a middle ground among these competing views instead of challenging them. He repudiates the Haeckel-to-Hitler thesis, but with the qualification that even if he did not lead the way to fascism, Haeckel represented a society that was moving in that direction. Di Gregorio expresses some skepticism of

Richards's account for dwelling on only the Romantic and idealistic aspects of Haeckel's thinking, but he himself makes comparably strong links between Haeckel and Goethe and between Haeckel and idealistic morphology. He criticizes Bowler for judging Haeckel by arbitrary standards and labeling him a pseudo-Darwinian, but ends up agreeing that "Haeckel was *not* a proper Darwinist. He simply thought he was one because he *needed* to"[53] (emphases in original). Di Gregorio repeats Gould's misleading claims about terminal addition and the deterministic nature of Haeckel's recapitulationism and asserts that Haeckel's interests and views on evolutionary morphology did not really derive from Darwin's, but only "converged" with them. To the extent that Di Gregorio's Haeckel was a morphological reformer at all, he was a very conservative one, whose Darwinism was only the means to a limited end, a "technical justification" for ridding morphology of its worst teleological and transcendental aspects while retaining as much as possible of its focus on types. Di Gregorio does say, however, that Haeckel's types were no longer transcendental but Cuvierian (as modified by Gegenbaur), so we get at least a partial correction of Russell and Richards.[54]

What Russell found most objectionable in Haeckel's treatment of evolutionary history was his conception of heredity as a conservative, constraining force. It seemed to him to doom the individual to repeat the past or, at best, to add a minor new wrinkle to an ongoing line of development. Ironically, when he turned to intellectual history, he had no qualms about treating "influence" in the same way, as a conservative, constraining force. He saddled evolutionary morphologists like Haeckel and Darwin with the ideas and assumptions of their predecessors among the transcendentalists. He let them only tinker with the definitions and concepts they inherited, not change them fundamentally.

Russell's thesis depends crucially on the method of tracing lines of influence, based on abstract associations among terms and concepts. So do the interpretations of Haeckel by Gould, Bowler, and Richards, even if the first two do allow Darwin to break from the past. And the same associations-game supports the Haeckel-to-Hitler and the more recent Darwin-to-Hitler theses, which trace National Socialism and the Holocaust back to intellectual influences from the nineteenth century and ascribe strong causal roles to similar long chains of associations.[55]

In what follows, I will be ascribing a much weaker sort of causality to intellectual influences. Early German morphologists did not stand for any one conception of form and development, but offered a variety of concepts and methods. Later morphologists and evolutionists were not passive recipients of influences from their German predecessors or from Darwin, but used the available intellectual resources selectively and also creatively when solving their own problems and building their own systems of thought and programs of research. They were not constrained to merely extend long lines of thought, as if by terminal addition. German Darwinism and evolutionary morphology had their origins not only in the past, but in Darwin, and in the creative minds of individual researchers like Bronn and Haeckel.

1

The Science of Life at the Turn of the Nineteenth Century

As we begin to reassess the continuities and discontinuities from pre- to post-Darwinian German biology and to situate Bronn and Haeckel in between, the first question is how to characterize the pre-Darwinian period. The following brief survey will establish a very different baseline from the one depicted by Russell and other authorities. It will show that there was more to early-nineteenth-century German biology than just morphology, more to morphology than just transcendentalism, and that even the transcendentalists were not such strict determinists and naive idealists and recapitulationists as previously supposed. All told, they made a variety of conceptual and methodological tools available for further development by Bronn and Haeckel.

The transcendental ideas, on which standard accounts of the period focus, did have a strong presence. With them came a desire to cut through the noise of the organic world, to look past the variation and see the essentials, the laws, and the underlying unity of nature's forms. But there was also a clear appreciation, even admiration, of nature's variety and creativity, and a consciousness of the need to explain where it came from. The German word *Mannigfaltigkeit* (in nineteenth-century orthography, also *Mannich-* or *Manchfaltigkeit*) was widely used to describe these phenomena of multiplicity, diversity, variation, or more literally "manifoldness." The wide usage of the term and discussion of the problem shows that German biologists did not have to wait for Darwin before they began to grapple with the problems of variation and of forms that defied categorization and typology. Morphologists could not realistically allow their laws of life to make straightforward predictions. The laws always took part in complex interactions, or were provided with exceptions and escape clauses that would make *Mannigfaltigkeit* emerge. Talk of law, determinism, and types predominated in morphological and systematic works of the period, but attention to the problem of *Mannigfaltigkeit* prevented the systems from ever becoming fully deterministic. In some cases, the veneer of law and order could be quite thin.

As natural history collections grew and diversified in the nineteenth century, and comparative studies of anatomy and embryology proliferated, nature's *Mannigfaltigkeit* became ever more evident and its explanation more urgent. Commitment to neat, abstract archetypes and linear scales of nature was continually weakened. Linear scales had to compete with Cuvierian classes or with branching or reticulate schemes. Concepts of type increasingly had to accommodate variations and intermediate or composite fossil forms. And embryos were not only seen as progressing up a scale of ever-more-perfect types, but also as differentiating or working out permutations of their type.

Karl Ernst von Baer is generally held to have overturned the idea of linear development along a scale of types and to have originated the more modern idea of differentiation and specialization within a type in 1828, but I shall argue that this is a myth. It has been perpetuated by Russell and many others, and it is based upon Baer's own self-aggrandizement at the expense of older colleagues, especially Johann Friedrich Meckel. Indeed much of Russell's caricature of transcendental morphology is taken over from Baer, and Richards's argument for an easy transition from recapitulation to transformation also depends crucially upon Baer's factitious account.

Thus, the purpose of this chapter is not so much to break the continuity between pre- and post-Darwinian evolution and morphology as it is to revise the picture of pre-Darwinian (and especially pre-Baerian) thought and to rediscover the wide range of problems and solutions that were explored. The new picture will reveal many more common interests between the Germans and Darwin than have been recognized before, while also allowing for differences in point of view that would play out in Bronn's translation and Haeckel's theoretical work.

Among the distinguishing features of German biology, I ascribe high importance to the scholarly ideals expressed in the German university reforms of the period, and the accompanying rhetoric of *Wissenschaft*, or pure and theory-oriented scholarship. Much more than their British counterparts, German life scientists concerned themselves with the status of their fields as scholarly pursuits and university-based disciplines. Even as they explored diverging prospects for a modern science of life, they also converged on certain *wissenschaftlich* standards and conventions. These included a strong emphasis on the unity of all the life-phenomena and all the sciences as well as on the orderliness of nature and the rule of law in both the organic and inorganic realms. As we shall see, the concern with establishing a *Wissenschaft* of life did not diminish for Bronn or even Haeckel, and it colored their expectations, interpretations, and applications of Darwinism.

Wissenschaft

One striking thing about German biological works from the early- to mid-nineteenth century is how frequently they called for new, more "philosophical" or "physiological" approaches to the study of life, approaches that would make biology into a *Wissenschaft*. At the very beginning of the nineteenth century, for example, Lorenz Oken announced the dawn of a new era in physiology and natural history, during which a new understanding of life was finally ready to emerge from the body of older descriptive work that had been its necessary precursor. Earlier researchers had to learn anatomy and be able to answer the what- and how-questions before Oken's generation could go on to answer the why: "This we know with certainty: that all the answers given before this century are false, through and through, if they are supposed to be physiological. The time had necessarily to be dedicated to pure anatomy."[1]

It was a refrain that could also be heard from the pioneering comparative embryologist Meckel in the 1810s and 1820s, who wrote of his desire to dress up the study of anatomy in "new, more *wissenschaftlich* and more dignified clothes,"[2] and of how anatomy had progressed from the mere description of individual animals to a new era of comparison and generalization.[3] And that refrain was repeated by many botanists and zoologists in the ensuing decades, each with his own vision of biological science to promote.

In other countries, too, leading naturalists characterized eighteenth-century approaches to life, no doubt unfairly, as mechanical exercises in collection, description, and classification, not up to the intellectual standards of other natural sciences. They called for analysis and explanation of all that had been collected and described: the multiplicity of forms, the patterns of resemblance among them, their geographic distribution, their development, and their complex functionality. In France, Jean-Baptiste de Lamarck gave a definition of "biology" that made no mention of description and classification, but emphasized the study of organization, complexity, movement, and change: "Biology: this is one of three divisions of terrestrial physics; it includes all which pertains to living bodies and particularly to their organization, their developmental processes, the structural complexity resulting from prolonged action of vital movements, the tendency to create special organs and to isolate them by focusing activity in a center, and so on."[4]

Similar calls for "philosophical" approaches to natural history were issued in Britain, mostly following French and German models, at first.[5] But as we shall see in chapter 3, natural theology also remained firmly entrenched in Britain long after it had become irrelevant for German academic biologists. As a result, Darwin's stratagems for countering (or co-opting) natural-theological reasoning went unappreciated by Bronn and Haeckel and seemed to them to require reinterpretation and reformulation.

For the German context at least, the story of this rebelliousness against descriptive natural history has been told before and linked to the first usages of the word "biology" for the emerging science of life.[6] The rhetoric of *Wissenschaft*, not only in biology but in all the natural- and social sciences and the humanistic disciplines (all *Wissenschaften* in German), has also been tied closely to the modernization of the German universities. The university reforms that began at Göttingen in the mid-1700s incorporated research into the mission of the university and the job of the professor, and—most important for present purposes—made room for natural history and other biological disciplines in the philosophical faculty, separate from theology and medicine (although some biological subjects, such as anatomy or medical botany, continued to be taught at the medical faculty).[7] The "Humboldtian reforms" of the early nineteenth century[8] stressed academic freedom, detached investigation, and the development, not just the transmission, of culture. The decentralized system also fostered competition among the German universities while letting multiple approaches flourish.[9]

The Rule of Law in Biology

One obvious system of method and explanation upon which to model a science of life was Newton's theory of gravitation. Its key features were quantification, explanations in terms of fixed and deterministic laws and forces, and the unification of certain disparate phenomena—from falling objects to planetary motion to ocean tides—by identifying them as effects of the same few causes. Conversely, those causes, or forces, could be known and studied through their effects.[10] However, it was not obvious how such an approach could be made to work in the more complex, variable, and apparently purposeful world of organic forms and processes. Most of the biological solutions to be discussed in this chapter adopted at least some features of Newton's system, but also had to balance these features with ways of producing *Mannigfaltigkeit*. Because the idea of biological laws and forces seems on its face to be too "vitalistic" to be admissible in scientific explanations, historians have given such laws and forces much too little serious treatment, even though they were among the few plausible, naturalistic explanations of life at the time,[11] and even though the physicalism and determinism were limited and qualified.

The application of Newtonian forces in biology was most conspicuous in the work of a group that Timothy Lenoir has dubbed the Göttingen school, whose leading lights were Johann Friedrich Blumenbach, Carl Friedrich Kielmeyer, and Johann Christian Reil.[12] Their approach emphasized the search for regular, predictable effects and quantifiable relationships that could be ascribed to the actions of forces, the most important of which was Blumenbach's *Bildungstrieb* (formative drive, urge, or tendency; *nisus formativus*). The *Reproductionskraft* and *Bildungskraft* (reproductive and formative forces) of Kielmeyer and Reil, respectively, caused comparable sets of effects, and I will treat them as variations on Blumenbach's theme.[13] Blumenbach focused on the sequence of forms through which the developing embryo passed on its way to maturity, as just the sort of predictable, repeatable phenomenon that could best be described as the effect of a force.

Blumenbach first presented his *Bildungstrieb* concept as an alternative to eighteenth-century theories of embryonic preformation, which had held that there was no way to account for the origin of organization and form in the embryo. Organization had to date back to the Creation, rather than being created anew in every generation. Embryos were in some sense preformed in the egg (or, in some versions, the sperm) and only had to grow and "evolve," in the word's original sense of unrolling or unfolding. The embryos were presumed to contain the rudimentary forms of the entire series of their potential descendants, nested each within its future parent, and Blumenbach seized upon this assumption as the fatal flaw of the entire system, for it was too constraining to do justice to life's adaptability and spontaneous creativity. How could such a predetermined chain of forms produce a hybrid, for example, unless divine providence anticipated the miscegenation and placed an appropriate embryo, intermediate in appearance between the parental forms, at just the right point in the series? How could it account for environmentally induced aberrations in form, unless, again, the preformed series anticipated every contingency?

The competing "epigenetic" school, which had each embryo develop its organization and form anew, could more easily account for environmental effects, hybridization, and other departures from the norm. Its main disadvantage, from the preformationist point of view, was that it needed to postulate mysterious, occult forces that would organize and form the embryo. The preformationists, needing no such forces, claimed to be the better mechanistic scientists.[14] The challenge for Blumenbach was to make a case for the existence of such forces, or at least for their legitimacy as explanatory devices. And he had to balance the rule of fixed laws with the flexibility and contingency of developmental outcomes, for without that flexibility, epigenetic laws themselves would have produced a predetermined sequence of forms and would have suffered from the same defects that Blumenbach found in preformationism. Blumenbach therefore had to have his biological forces interact with the physical environment in complex and unpredictable ways.

In making his case, Blumenbach argued that there were three major formative processes similar enough to be subsumed under the effects of one and the same cause: something caused the developing embryo to go through a predictable series of forms on the way to maturity (development), maintained the mature form by assimilating new materials (nutrition), and restored the original form in case of injury (regeneration). Putting aside the question of what it actually was, Blumenbach named the cause of these effects the *Bildungstrieb*. It was generated somehow by organic substances under special conditions, and Blumenbach supposed that every species had its own *Bildungstrieb* that produced and maintained its characteristic form.[15]

According to Blumenbach, developmental outcomes were not fully determined by this force. Environmental influences could derail the *Bildungstrieb* and make it produce aberrant individuals, with novel forms and even new adaptations. Not only that, but these changes could sometimes become permanent. The altered *Bildungstrieb* could be generated again in the offspring of the modified individuals, giving rise to new varieties. Blumenbach did not elaborate much on the evolutionary implications of his theory, and it might not have been meant to produce changes above the variety level, but still it included at least some historical change among the effects of the *Bildungstrieb*. Blumenbach thus pointed the way toward a unification of development and evolution under the rule of physicalistic laws and forces.

A different account of how biological forces could generate variety and complexity can be found in Kielmeyer, an elusive author who was widely cited by other biologists as an inspiration but who published very little. He has only gradually gained the attention of historians, who are beginning to reanalyze his one well-known publication and bring his unpublished lecture notes and manuscripts to light. That publication, which will be analyzed here, originated as a lecture, delivered in 1793 and titled "Ueber die Verhältniße der organischen Kräfte unter einander in der Reihe der verschiedenen Organisationen, die Gesetze und Folgen dieser Verhältniße" (On the relationships of the organic forces to one another in the series of different organizations, and the laws and consequences of these relationships).[16]

Among other things, the lecture outlined a system of physicalistic forces in biology. Kielmeyer recognized five classes of effects that needed to be unified and explained: sensitivity, irritability, reproduction, secretion, and propulsion, of which reproduction comprised the formative processes.[17] Applying Newton's rule that like effects be ascribed to the same cause, he inferred that there was one cause for each of his classes of phenomena, and in the absence of further information about their nature, he referred to them, provisionally, as "forces." Finally, in the manner of Newton declining to feign hypotheses about gravitation, Kielmeyer left aside questions of what those forces were and what made them work, and sought instead to characterize their behavior, if possible by means of mathematical laws. He carried his physicalistic methodology further than Blumenbach by suggesting ways of measuring effects quantitatively in different organisms and under different conditions, and he gave preliminary formulations of laws governing each force. Among the measurements he proposed were the frequency with which a force acted, the range or duration of its effects, and the resistance it could overcome.[18]

Based on such measurements, Kielmeyer found, for example, that the reproductive force obeyed several laws, among them this one, relating the number of offspring produced to their initial size, complexity, and gestation period: "The more the reproductive force expresses itself . . . in the number of new individuals, the less is the mass of the body of the new individuals; the simpler are the bodies constructed with which they appear; the shorter the time taken for their formation in the bodies of the parents."[19] Laws of this sort described relationships and trade-offs between variations, but did not determine or predict which variations would appear or how an organism would turn out.

A third variation on physicalistic forces was due to Reil, another author associated with Blumenbach and the Göttingen school. In a famous 1796 paper on life forces, Reil emphasized the unity of nature and the fundamental similarity of biological, chemical, and physical forces. The many forces of nature could be arranged in a hierarchy of specificity and selectivity. The gravitational force was least specific, for it attracted all objects to one another. The magnetic force was more selective, attracting and repelling only certain metals. The forces of chemical affinity were more selective still, with regard not only to the particular substances they combined and transformed, but also the proportions of those substances. Finally, the *Lebenskräfte* (life forces) were the most discriminating, affecting only organic materials and only when they were present in the right *Form und Mischung* (form and mixture). Among the life forces Reil counted a *Bildungstrieb* (or as he preferred to call it, *Bildungskraft*) like Blumenbach's, and various physiological forces comparable to Kielmeyer's, but even more specialized. Some occurred only in individual organs and in individual species. For example, the forces that made rational thought possible acted only among the substances that made up the human brain.[20] Reil thus subsumed all natural phenomena within a single system of laws and forces, giving unity to all the sciences and leaving no protected space for biological change, outside the reach of physical forces.

Yet another important physicalistic approach to organic nature was that of Alexander von Humboldt, who combined the holism and aesthetic sensibility of the German Romantic with a zeal for empirical description and quantitative measurement and correlation. His main interest was in physical and biological phenomena that interacted on a global scale, and he traveled widely, making maps and statistical tables to document changes in climate, flora, geomagnetism, and other variables over great distances. Species counts within taxonomic categories or ratios between such counts were used, by Humboldt and others, to characterize a regional flora or fauna or to quantify the biological effects of different environmental variables. Laws could then be derived from numerical patterns in the tables or on the maps.[21]

In the spirit of Humboldtian quantification, many other biogeographers compiled extensive tables of species counts and ratios in different categories and locations (*Tabellenstatistik*). They also raised questions about geographic distribution that seemed to require historical if not evolutionary answers—questions about why the same environmental conditions might be associated in different places with different species. Similar numerical methods entered into other fields, such as medicine, where patterns of occurrence of diseases were related to physical and geographic variables,[22] and paleontology, where they were used to characterize the flora and fauna of successive geological strata.[23] Thus, Humboldtian methods broadened the scope of physicalistic biology in space and time, and pointed toward a quantitative, *wissenschaftlich* study of geological history and the fossil record as well. Bronn and Darwin were both direct beneficiaries of Humboldtian methodology and thought.

Kant, Kielmeyer, and Complexity

One of the central problems with physicalistic laws and forces was how to use them to account for complex and purposeful activity and adaptation. The problem was especially acute because the philosopher Immanuel Kant had argued that it could not be done. The complex structure, the mutual adaptation of parts, and the harmonious functioning of organisms, he argued, could not be understood, even in principle, without reference to ideas, abstract archetypes, and purposes, and these had no place within a Newtonian paradigm of quantitative, mechanistic, law-bound science. Therefore, Kant argued, no Newton could ever arise and explain so much as the generation of a blade of grass in terms of the unintentional action of laws of nature, and life could never be approached in a properly scientific manner.[24] The best that naturalists could do was devise their explanations of life "as if" the organic world were teleological, and to invoke ideas, archetypes, and purposes in order to explain form and function. However, the usage had to be recognized as merely heuristic, in lieu of the proper physicalistic explanation that would always remain out of reach.

Ever since, historians of biology have been judging German biologists by their adherence to Kant's strictures on teleology. Were they good Kantians, whose talk of purposes or transcendental ideas was only meant heuristically? Or did they attend only to the part of Kant's argument that said that ideas and purposes were indispensable in biological explanations, and then go ahead and use them unabashedly and try to understand them? The latter was the course taken by diverse thinkers generally categorized as Romantic *Naturphilosophen*, idealists, or transcendentalists, who dominate Russell's account of pre-Darwinian morphology. But in practice, it is not always easy to tell who followed the former course, or when a given author's usages were meant only heuristically and when the teleology was a constitutive feature of a theory. The authors of the time had good reasons to be evasive on these points, for they would not have wanted to contradict the high authority of Kant, by being either too teleological or not teleological enough.

Authors like Blumenbach and Kielmeyer are therefore best viewed as neither obeying nor defying Kant's strictures, but as testing the limits of their physicalistic methods and interpretations. How far could they get before they were forced to invoke transcendental ideas and purposes? As biologists, rather than philosophers, they found it worthwhile to try to elucidate causal laws and forces, measure and quantify effects, and analyze the behavior of complex, purposeful systems, even if they had to concede to Kant at some point that ideas and purposes appeared to be involved.

In any case, Kant gave Blumenbach's approach an influential endorsement, under the assumption that Blumenbach actually meant the goal-directedness of the *Bildungstrieb* to be merely heuristic, as opposed to a constitutive property of the force.[25] Lenoir reads Blumenbach, Kielmeyer, Reil, and Meckel the same way, and frees them from suspicions that they belonged in the transcendentalist camp, where Russell originally had them (or at least Kielmeyer and Meckel).[26] Coleman has done Kielmeyer the same favor.[27] On the other hand, Richards is trying to win them all back to transcendentalism. He needs them, especially Kielmeyer, in his chain of influences that bind Darwin (and Haeckel), through recapitulation theory, to German Romanticism and teleology.[28] He argues that Blumenbach creatively "misunderstood" Kant to be endorsing his use of teleology (rather than just his application of laws and forces), a usage that Richards reads as more than just heuristic. He also gives a stilted reading of Kielmeyer as a teleologist and recapitulationist that does not stand up to scrutiny.[29]

That so many conflicting interpretations could arise seems to me to confirm that the German authors did not lay out their views on the nature, cause, and directionality of their developmental laws and progressions with great clarity and precision. Indeed they had little choice but to equivocate. If Blumenbach, for example, had accepted Kant's argument fully, he would have had either to concede the inadequacy of his physicalistic approach or to invoke (even if merely heuristically) the kinds of nonmaterial ideas and essences that he had already explicitly rejected in his 1781 monograph (which predated Kant's 1790 *Critique of Judgment*). Neither option could have been attractive, so it is hardly surprising that in later versions of his argument he played up the resemblance of the *Bildungstrieb* to a Newtonian force, and implied that Newton gave him the license to describe it and seek the laws that govern its actions, without hypothesizing about its nature and causes.[30]

One cannot expect Kielmeyer, any more than Blumenbach, to have embraced either of those options. Instead, he couched his ideas in

Kantian as well as Newtonian language as much as possible, but cultivated ambiguity on the finer philosophical points. But there was also more to his approach than the mere derivation of physicalistic laws and forces. He addressed the problem of complexity head-on and argued that an organism was a system of interacting parts whose overall behavior was still amenable to scientific study, even if detailed knowledge of the parts and their interactions and purposes was out of reach. The patterns and sequences of the system's activities could be described, quantified, and rendered predictable. This consideration of temporal patterns led to his idea of a parallelism between embryonic development and the history of life as revealed in the fossil record.

In his lecture on the organic forces, Kielmeyer went so far with Kant as to say that organic nature *appeared* to us as wondrous and purposeful, but he stopped short of saying either that it actually was so, or that we were merely obliged to think it so. Kielmeyer argued in one very long paragraph that the complexity of the organic world was indeed amenable to scientific study, but that it was in no danger of being demystified by such study. It was still capable of instilling in us a sense of awe, akin to what Kant said we felt when we contemplated the cosmos: "Although there are no masses, volumes and distances here [i.e., in organic nature] like those through which nature in the heavens convinces us of her greatness, there are things of another sort. These things speak no less penetratingly for her greatness [*Größe*], if we can come down to their level, in a way, and allow the multiplicity [*Vielheit*], diversity [*Mannichfaltigkeit*] and harmony of effects in a small space and over short periods of time to have a voice in the decision about the greatness of an object, and if we listen to that voice with a little patience."[31]

Most impressive of all, to Kielmeyer, was the dynamism that followed from the multiplicity, diversity, and harmony of these complex systems of interacting parts. What made it dynamic was that every part functioned as both cause and effect of the others and of the organism as a whole: "In the changes that it experiences at every moment, each organ is adjusted to the changes of all other organs, and these are united in a system of simultaneous and successive changes, in such a way that every change, in our manner of speaking, is alternately [*wechselweise*] cause and effect of the others." One change led to other changes elsewhere in the system, and these to others still, and as a result, the organism changed and developed continually: "And at every point in this temporal trajectory, the system of effects that we call its life, and the system of organs that makes up its body, modifies itself, one stage coming out of another as if proceeding from its cause. In each individual, childhood, youth, old age, and death in turn extend a hand to one another."[32]

The dynamism was not limited to the individual, for the individual was embedded in the larger systems of its genealogical lineage, its species, and the collective of interacting species. Human society, too, was a dynamic system, moving along a developmental trajectory (*Entwicklungsbahn*). All living things were part of a great dynamic system. They were "linked together in the life of the great machine of the organic world, and this machine, too, seems to be advancing along a developmental trajectory."[33]

The machine metaphor suggested a mechanistic and reductive approach to life, but in the same long paragraph Kielmeyer also toyed with the idea that there might still have been an ultimate purpose for the machine to serve:

Suppose that nature had had no purpose in this artful placement of phenomena in time, one after the other and one beside the other, and that those effects and consequences were not aims she had wanted to achieve. Suppose that it would be empty dreaming if we wanted to surmise that there were higher and to-us-unknown purposes in this. Then we will still have to admit that this concatenation of effect and cause, in most cases, looks, at the same time, like a concatenation of means and end in our human affairs, and we will even find it very helpful to our reasoning to assume such a concatenation. And with that, we will have to admit now at the end, at least, that nature would be no less capable here than in the heavens to convince us of the truth of the point from which I started out.[34]

Thus the end of Kielmeyer's long paragraph links up with its beginning, with an appreciation of nature's greatness, which could be appreciated just as well in the organic world as in astronomy.[35]

Aesthetically, according to Kielmeyer, nothing was lost under the assumption of purposelessness. The complexity, diversity, and harmony of organic nature impressed and inspired us as much as the vastness of the heavens, regardless of what we might conclude about teleology. And scientifically, we could put the question of teleology aside and get on with the analysis of life processes. Even if we could never get at the inner workings of the organic machine, it still offered patterns of behavior for us to describe, quantify, and bring under the rule of law.

Types of Types, and Goals of Development

In addition to studying the process of change and the quantifiable laws governing it, morphologists also needed scientific tools for describing and comparing the intermediate forms and the endpoints of the process.

The definition of taxonomic groups and the search for a natural system of classification required such tools. The very soul of natural history, as Darwin would later call it,[36] was the study of body plans, or general patterns of organization, and especially the phenomenon of "unity of type," by which was meant the ability of members of a taxonomic group to use the same underlying body plan for many different ways of life.

It was long assumed that, before Darwin, concepts of type were bound up with abstract forms and essences, and that the persistent idol of "essentialism," in a number of different guises, impeded the acceptance of Darwinism and later of the modern evolutionary synthesis. This "essentialism story" derives mainly from Ernst Mayr, but, ironically, it fails to take into account the great variation among type concepts, and it underestimates the pre-Darwinian interest in the diversity of life forms, as Mary P. Winsor and other authors have been arguing.[37]

As Paul Farber has shown, the type concept, for all its evident practical and theoretical importance, was defined only very loosely and interpreted variously in the nineteenth century. And according to Gordon McOuat, social mechanisms and conventions were more important for maintaining workable taxonomic systems than were concepts of species or type, on which no consensus was reachable. For purposes of classification, an actual specimen might have served as the type of its species. It did not have to be average or even "typical" in appearance; it only had to be a fixed reference point for the name and the formal description of the species, which could include information about the range of deviations from the type specimen. There could also be a hierarchy of types: an individual provided the type of the species, a species the type of the genus, a genus the type of the family, and so on.[38]

In comparative morphology the type tended to be not a particular specimen but an abstraction, summarizing the distinguishing features of a species or higher group. Sketched in diagrammatic form, it was at the very least a convenient tool for characterizing a group, but transcendental morphologists ascribed an independent existence to it, as a kind of Platonic form or divine idea. On this view, unity of type obtained because group members were all imperfect instantiations of the same transcendent idea or because the idea somehow caused or guided development.

Like those used in classification, these transcendental types could also be organized hierarchically, reflecting the overall unity of living things and the plan of nature. Johann Wolfgang von Goethe, the poet-naturalist, gave morphology its conception of the *Urpflanze* or plant archetype, which contained the sum total of all the basic parts of all plants. Actual plant species realized selections from and variations on the archetype's parts.[39] In zoology, Oken's system of *Naturphilosophie* used the human body as the highest, most perfect, all-inclusive type, and all the lower groups and forms were derived from subsets of its parts and faculties.[40]

As Nicolaas Rupke explains, such methods of deriving the lower forms "by subtraction" from the perfect, all-inclusive archetype was one of two main options for building a hierarchical, transcendental system.[41] The other option was to begin at the bottom and include only the common features of the group within the type. Higher types and instantiations were derived "by addition" of new features and modifications. Rupke's prime example of such a minimalist archetype is that of Richard Owen, the British zoologist and comparative anatomist. Owen's vertebrate archetype was little more than a string of vertebrae, including their various projections and extensions, and it resembled the lowest fish. To this common denominator, higher groups added, for example, limbs and lungs.

It should be noted that there were also German minimalists such as Carl Gustav Carus, whose vertebrate archetype might have served as a model for Owen's. And Oken, too, sometimes derived his system from the bottom up as well as the top down. In his "mathesis," Oken derived all of nature, in Pythagorean style, from the number zero, and elsewhere he described life emerging and being built up from primeval slime bubbles.[42] Thus the hierarchy could be a two-way street. Even Owen's minimalist system had a maximalist "top." Owen located his archetypes not in some Platonic realm, but as ideas in God's mind, where all the potential additions and modifications were present in divine forethought.

There were further differences among transcendentalists over what role archetypes played in development or evolutionary change. In systems like Oken's, every stage of embryonic development corresponded to a different idea, which was a component of the overall idea of nature, and the embryo instantiated one idea after another. In Baer's system, as will be discussed in greater detail below, the embryo never changed ideas during the course of development, but was guided by a single one, which it came to embody gradually by degrees of differentiation and specialization. The idea of the adult played an active role in shaping the organism and ironing out environmentally induced developmental variations. For Blumenbach and others of the Göttingen school, the idea of the adult form was somehow built into the *Bildungstrieb*; it did not watch over development, as if from afar in a transcendental realm, to correct for environmental influences. The *Bildungstrieb*, along with its built-in ideas, interacted with the environment and could be changed by it. Christian Heinrich Pander shifted the balance even farther to the environmental side by arguing that idealized and unchangeable forms failed utterly to explain the "manifold diversity [*manchfaltige Verschiedenheit*]" of organisms. Not only that, but they also defeated the whole purpose of doing comparative work, which was to account for the variety as well as the unity of life. In fact, Pander placed so much emphasis on the role of the environment, especially food, that he was prepared to give up abstract forms altogether: "Because, however, we also recognize conditions and relationships among all the material constituents of organisms: it follows that in order to explain organic activity as a manifestation of life, the assumption of a merely non-material essence that expresses itself in the *Bildungstrieb* is neither necessary nor sufficient."[43]

In addition to actual specimens or transcendental forms, types could be understood as blueprints or functional designs, as they were in the system of the French zoologist and paleontologist Georges Cuvier. Cuvier argued that the animal needed to function as an integrated whole, and that hence its form was determined not by an abstract idea, but by practical considerations. Unity of type was explicable because there was only a finite number of basic designs that actually worked. Nature had to reuse the same ones in many organisms. These functional considerations also placed limits on the unity of the natural system as a whole, since Cuvier held that no common plan could be imagined for all animals. At the top of his taxonomic hierarchy, therefore, Cuvier had not one, but four main types, classes, or *embranchements*. Each of the four—the vertebrates, mollusks, articulates (worms and arthropods), and radiates (coelenterates and echinoderms)— was based on a general body plan so different from the others as to be completely incommensurable with them. There was no unity among these classes and no scale of progress from "lower" to "higher."[44]

Cuvier's functional type was based on reasoning similar to Kielmeyer's in his conception of the organism as a complex system, perhaps reflecting the close association between the two men early in their careers. Both emphasized the ways in which the parts of the organism interacted in complex and harmonious ways; only Kielmeyer, though, described the interactions as dynamic, with any change in one part initiating a cascade of consequences throughout the system, and causing development and possibly evolution. In contrast, Cuvier placed stricter limits on change within the system, envisioning such a delicate balance that change in a single part was more likely to cause breakdown than evolution. Cuvier's solution preserved the unity and coherence of each major type, but at the expense of continuity between types and the overall unity of nature.

Scales of Perfection and Recapitulation

Cuvier's discontinuous classes represented a break with a long tradition of arranging organic forms on a continuous chain of being or scale of nature, extending from the lowliest plants through the increasingly perfect animals, humans, and in the more philosophical or theological versions, the angels and God as well. But other biologists were also moving away from a strictly linear and continuous arrangement, and they were revising the concept of "perfection" to make it more biological than moral, spiritual, or anthropocentric. In the late eighteenth and early nineteenth centuries, for example, authors such as Charles Bonnet, Georges Buffon, and Goethe began to enunciate functional and morphological criteria by which to measure perfection and arrange forms on the scale. They described higher forms as *adding* new functions or faculties, concentrating or *condensing* functions into fewer repeated segments or organs, or becoming *differentiated* into more specialized organs and parts.[45]

Kielmeyer is generally considered the first to have transferred some of these criteria of progress and perfection from the scale of nature to the development of the embryo. In his account, progress took the form of addition of faculties. He noted that in the embryo, his five physiological forces became active one at a time, in a predictable sequence, and each one endowed the embryo with another set of functions and faculties. First the productive force set development into motion. Then the force of irritability was activated, providing the embryo with the ability to respond to stimuli and to contract certain of its tissues. Later, the forces of secretion and propulsion came into play, and last, but highest of them all, sensibility, and with it, the faculties for perception and mental activity. Comparable patterns of addition or changing dominance of forces could be identified in individual organs, for example the uterus, which changed its state at appropriate times from predominantly secretory to productive to irritable and contractile. Kielmeyer was outlining a general framework for the analysis of biological change.

Changes along the scale of nature were amenable to the same sort of analysis. Contemplating the scale from the bottom up, Kielmeyer again saw forces and faculties being added in a sequence comparable to that observed in the embryo. The productive was common to all levels, the sensible was added only at the higher levels, and the other forces and faculties were distributed in various ways in between, though *not* in a consistent or uniformly progressive pattern, and *not* in the very same sequence observed in embryos. There were complex trade-offs and laws of compensation that allowed for wide fluctuations in the activity of every force at every level in the scale.

Kielmeyer argued, further, that in the actual history of life on Earth, as revealed in the fossil record, the same forces and their associated faculties became active in the same kinds of additive patterns and according to the same progressive and compensatory laws that were observed in the scale of nature and the embryo. He concluded that the same forces and the same associated laws were at work in all three sequences—the embryonic, the systematic, and the paleontological.[46] On the history of life, Kielmeyer was short on detail, but he seemed to be espousing a theory of actual species transformation, as opposed to a succession of separate creations or spontaneous generations. An analogy he drew between the origin of new species and the metamorphosis of caterpillars to butterflies suggested transformation rather than succession, perhaps in a saltational mode with lineages adding faculties all at once.[47]

These observations of Kielmeyer's are usually cited as the starting point for the doctrine of recapitulation or parallelism, which linked the embryonic sequence to the scale of nature, the history of life, or, as in Kielmeyer's case, both. But note that Kielmeyer's parallels were not morphological ones, among sequences of increasingly perfect or specialized forms. They were very inexact parallels—more like analogies—among three patterns of activity of the physiological forces and faculties. Progress occurred not by change in form, but by addition of new faculties or increasing predominance of an existing one. Kielmeyer made no claim that any embryonic form resembled or corresponded to any historical adult or lower taxonomic type, and he made no provision for deducing ancestry or evolutionary relationships from embryological evidence.[48] Kielmeyer's metaphorical and conceptual links between embryology and evolution were based on the perception that they were kindred developmental processes, governed by the same laws and forces, which became active in similar temporal patterns.

The best-known early recapitulationists, such as Meckel, Oken, Friedrich Tiedemann, or Gottfried Reinhold Treviranus in Germany, and Étienne Serres in France, built upon Kielmeyer's foundation by introducing the morphological parallels that E. S. Russell characterized as the Meckel-Serres law. However, even in their work, morphological parallels were drawn only between embryonic stages and forms on the scale of nature. There was no recapitulation of ancestral adult forms, not even among those authors who did allow for evolutionary transformation. The transformationists were more likely to draw parallels in the general direction of change, the laws and forces involved, or the forms of single organs. Or morphological parallels were dispensed with entirely, and the entire plant or animal kingdom or regional fossil record was described as a developing superorganism, with paleontological species-assemblages as its embryonic stages. For example, Tiedemann wrote, "Just as each individual begins with the simplest formation and during its metamorphosis becomes more evolved [*entwickelt*] and developed, so the entire animal organism [i.e., animal kingdom as superorganism] seems to have begun its evolution [*Entwickelung*] with the simplest animal forms."[49]

Similarly, Treviranus consistently treated all of organic nature as a single living organism, and he ascribed law-abiding developmental tendencies to it: "We will therefore, first, view living nature as a whole that has always been, still is, and ever will be involved in making steady transformations [*Umwandlungen*]; but also, second, assume in these transformations a fixed, lawful path."[50]

Oken made the strongest claims of all for morphological parallels, but his were only between the embryo and the animal scale, and had nothing to do with ancestry. Also, even he never described an embryo as *identical* to a lower adult, as Russell suggests all recapitulationists did. Oken matched embryonic stages with animal groups on the scale of nature by their overall levels of organization or by a single representative common feature or faculty. There was no embryonic recapitulation of the entire form of any lower adult, only an approximate, metaphorical, or even synecdochic parallel. He was aiming not for exact examples, but only sufficiently suggestive ones to convince readers that there was some kind of relationship between embryology and systematics: "Even should the account of these parallels not be everywhere correct: it follows clearly enough that a complete [*vollkommener*] parallelism occurs between the *Entwicklung* of the fetus and that of the animal kingdom."[51]

Meckel on Progress and Diversification

Meckel's system of recapitulational embryology was by far the most influential. It was also the most thoroughly documented, based as it was on detailed comparative work on many different species, not just vertebrates, but representatives of all the Cuvierian classes, as well as normal and abnormal human fetuses. It also contained thoughtful analyses of the direction of development, the nature of differentiation and complexity, and the possible history of evolutionary progress.

Although Lenoir does not count Meckel as one of the Göttingen school, he was closely associated with them, personally and intellectually. He studied with Reil between 1798 and 1801, soon after the latter published his analysis of *Lebenskraft*, and he was a long-time colleague of Reil's at the University of Halle. He also spent two semesters with Blumenbach at Göttingen, and later worked with Cuvier in Paris, where he made the acquaintance of Humboldt and probably Lamarck and Geoffroy as well.[52] In his own work, however, Meckel no longer felt the same need as the older scientists of the Göttingen school to establish the applicability of physicalistic laws and forces in morphology, systematics, and the history of life. His project was to work out the details of what the laws required. He also confronted the tension between the rule of simple, deterministic laws and the diversity and variation of life, and he sought ways to make the same laws generate multiple outcomes depending upon initial conditions and materials, and the speed, timing, and coordination of their effects.

Meckel exhibited little patience for philosophy or high abstractions, seldom saying much about the problems of teleology or the nature of the developmental forces. He seemed to take Reil's general ideal of a *Bildungskraft* for granted, which he mentioned on rare occasion and described as an organic analogue to physical forces such as electricity or magnetism.[53] He also paid no more than the merest lip-service to teleology. He wrote that there were two distinct ways of looking at organisms, one physicalistic and one teleological: "The animal form can be viewed either as a matter in and of itself and in relation to the physical force to which it immediately owes its existence, or in relation to the purpose to be achieved through it, and to the ideal [*geistig*] force that, as its creator, underlies it."[54]

The latter, idealistic view was of little interest to Meckel. With appropriate Kantian hedging, he allowed that the complex organization of animals "cannot be thought of" without teleology as an ideal or mental principle (*geistiges Princip*) and ultimate cause. A few isolated statements like this have earned him a reputation for transcendentalism, even though, in practice, teleology and transcendental archetypes never entered into any particular analysis or argument of Meckel's. He never used them to explain development or to define its goals or endpoints. He referred to taxonomic groups as collections (*Sammlungen*) of individuals, not reifications of archetypes, and he made the boundaries between groups quite fluid, allowing for species transformation but not necessarily common ancestry. He posited multiple spontaneous-generation events as historical starting points, and many possible paths of development and transformation in every lineage or taxonomic group.

After playing homage to teleology, Meckel hurried on to discuss the twin biological problems of diversity and unity, or, as he called them, *Mannichfaltigkeit* and *Reduction*: "Since the description of any single organ is a more-or-less clear proof of this law [of purposefulness], a further treatment of the same would be fully superfluous here; in contrast, the laws of *Mannichfaltigkeit* and *Reduction* deserve closer consideration."[55] Like Kielmeyer and Blumenbach, he pushed a program of developmental law as far as it would take him, and held the transcendental in reserve for only the most intractable problems and for ultimate causes.

Of his dual problem-areas, *Reduction* had to explain the appearance of unity amid diversity, the consistent patterns and apparently deterministic laws of development that held across all the animal groups and all developmental stages. *Mannichfaltigkeit* had to do the converse: provide the causes of difference and variability that emerge despite the unity and predictability of the developmental patterns. Contrary to his reputation for finding strictly linear, deterministic patterns of change, Meckel sought a balance between these two projects, which he placed almost in dialectical opposition.

For Meckel, the unity of the animal kingdom was evident in the earliest stages of every embryo, which were virtually indistinguishable from one another, even if they came from different Cuvierian classes. Even though embryos developed specific differences very quickly, the unity of all the animal groups could still be seen in later embryonic stages, in the patterns of developmental progress, which ran in parallel in all species. These were repeating patterns, which began with the addition of a new organ or organ system. The new organ was initially undifferentiated and diffused throughout much of the body. Later it began to "condense" around multiple centers or segments. Then these centers differentiated from each other into specialized structures and merged with other such structures to produce complex organs. Parallel sequences of addition, condensation, differentiation, and aggregation were repeated by every organ on both the temporal scale of individual embryonic development and on the scale of morphological perfection of animal groups.[56]

Meckel invariably analyzed these developmental patterns organ by organ. For example, when the nerve chord made its first appearance in the embryo, it was long and approximately uniform. It developed condensations in the form of rows of swellings, most of which went on to become ganglia. The swellings at the head end merged into a brain and continued to differentiate into more and more complex convolutions.[57]

Not every embryo ran through the whole sequence, of course. Embryos of lower organisms stopped before developing a complex brain or perhaps any brain at all. Pathological cases in higher organisms might also not complete the sequence. Higher organisms might also skip stages. Generally they had to run through the sequence at a faster tempo than lower organisms, because they had more, and more highly differentiated, organs and therefore farther to go in the time available for development. But the overall pattern held across the animal kingdom, and the result was Meckel's law of parallelism, which was actually this: "And so the same organism, from its first appearance to a certain stage of its existence, runs through the most important levels of which the series of organic forms consists."[58] It was a parallelism of levels of organization, not of precisely repeated forms.

The simplest way for nature to produce *Mannichfaltigkeit* out of this unified, law-abiding system was to have different species develop at different rates and finish at different points on the scale. Meckel worked out the development of the heart in special detail, describing its origin in all animals as a tube-like structure, and its subsequent differentiation into an auricle and a ventricle. This stage was as far as many embryos, such as those of clams or snails, ever got. For them, it was the "permanent" form of the heart, for it persisted unchanged into adulthood. Fish and lower amphibians added an aortic bulb to the auricle and ventricle. In higher vertebrates the auricle differentiated into two, and then the ventricle began to divide as well. Reptiles varied in the degree to which the ventricle was divided. In birds and mammals the interventricular septum was complete, except in pathological cases (heart murmurs), which Meckel explained as incomplete or arrested development of the heart, or at least of the septum.[59] Even if all the organs condensed and differentiated in synchrony with the heart, Meckel could generate a good deal of diversity or *Mannichfaltigkeit* along the linear scale of progress, but other possibilities were available on his theory.

Variations among organs in the tempo of their development and their stopping points provided him further means of creating *Mannichfaltigkeit* among species and among individuals of the same species. Individual organs or organ systems might have been on linear tracks, but they were independent from one another. If one organ finished developing at a low point on the scale, other organs could still continue to appear, condense, and differentiate.[60] The result was that many more different configurations could be produced than if all the organs had to develop in synchrony. Depending on its membership in a class, order, genus, species, and variety, and also on its sex and its individual peculiarities or pathologies, each embryo's overall development could run a unique course.

The Direction of Development

In what might have been a small concession to Cuvier and an anticipation of Baer, Meckel distinguished between developmental differences of degree and of kind ("*sowohl dem Grade als der Art nach*").[61] The degree or grade of development (*Grad der Ausbildung*, or sometimes *Grad der Differenzirung*)[62] was his measure of how far the process of condensation and differentiation had gone. The concept of a degree of development allowed unlike organs or body plans to be compared with one another and ranked on a scale of how far the process of condensation and differentiation had progressed.

The organ-level applicability of Meckel's law was the feature that is now most often overlooked and misrepresented. Meckel never made the ridiculous claim, attributed to him by Russell, that "the lower animals are the embryos of the higher."[63] He even anticipated being misunderstood on this point and made an effort to clarify his view. After describing a headless aborted fetus in terms of slowed or arrested development in the region that should have produced the head, he cautioned that the arrest was only regional. The embryo as a whole was not a lower organism:

> It should not need to be mentioned that I did not mean to say: there is an early state in which it is, e.g., normal to have strongly developed lower extremities, but no upper ones and no head, etc. My opinion can only be that there is a state in the development of the fetus, where the organs that one finds to be absent or abnormally developed in such monstrosities really are behaving according to the norm. The monstrosity then originates through the circumstance that these organs alone do not continue to develop in the usual manner, while others more or less run through all of their stages.[64]

The same went for the nonpathological cases. In a typical formulation of his law of parallelism, Meckel concluded that "The development of the individual organism occurs according to the same laws as the entire animal series, i.e., the higher animal in its development, in its *essentials*, runs through the permanent stages of those below it, allowing both temporal differences and class-differences to be reduced to one another" (emphasis in original).[65]

As in the case of the headless monstrosity, Meckel was not saying that the embryo as a whole ever took on the adult form of a lower species, only that its individual organs appeared, differentiated, and condensed in patterns that ran in parallel to progress up the animal scale. When Meckel did speak of the embryo as a whole reaching a certain level, that was always a judgment about its overall condition, based on the states of the organ or organs that the naturalist considered most relevant for comparative purposes.

For example, when the heart of the human embryo differentiated from the two- to the three-chambered stage, Meckel would have said it had reached the level of complexity of the class of reptiles, in which the three-chambered stage is usually as far as the heart ever progresses. In Meckel's terminology, the three-chambered stage was the abiding or permanent [*bleibend*] one in reptiles. Other organs or systems might develop at different rates and be at different levels of complexity from that of the heart in the same embryo: "As long as a certain organ has a certain form that belongs abidingly to a lower class, then the embryo of the higher animal apparently belongs to the lower class *with respect to this organ.*"[66]

Meckel was well aware of Cuvier's critique of the continuous, linear scale of nature and the difficulty of making morphological comparisons across the major classes. But he still thought it possible to make valid comparisons between Cuvierian classes, especially if one focused on the least-differentiated organs. Where specific organs could not be matched up and compared, he could still judge relative height on the scale, based on the number of organ systems and the overall degree of differentiation and condensation in them.[67] For example, segmented worms were higher than unsegmented ones because their bodies had condensed into more parts and the parts had begun to differentiate. Insects were higher than either the segmented worms or their own wormlike larvae because so many of their segments had merged and specialized. Spiders were higher than insects, because whereas insects breathed through spiracles and tracheae, arachnids had spiracles and book lungs, which consisted of fewer repeating parts.

Perhaps also in deference to Cuvier, Meckel did not cling dogmatically to a linear scale, but allowed for other possibilities. In particular, he considered a bifurcating arrangement as an alternative. One line went from the segmented worms through the insects, arachnids, and crustaceans to the bony fish; the other from the unsegmented worms through the mollusks and cephalopods to the cartilaginous fish and on to the amphibians and the rest of the vertebrates. The barnacles bridged the two lines in the middle, between the crustaceans and the mollusks. Each line progressed in its own way, by adding, condensing, and differentiating its own sequence of organs.

Up to this point, Meckel's parallels were strictly between embryos and the animal scale, but he also considered species transformation likely, and desired to extend his system to historical patterns of change. He accepted many of the ideas of Blumenbach and Lamarck about the inheritance of environmental effects, the effects of use and disuse of organs or parts, and how these effects could increase the *Mannichfaltigkeit* of animal nature. But he found that they did not sufficiently constrain and regulate change in such a way as to maintain unity. A form could not be allowed to change into just anything. Evolutionary change had to parallel the embryological pattern of addition, condensation, and differentiation of organs or systems. New forms could branch off from a lineage at any point, if an individual's development (in any organ or organs) were either inhibited below the species' norm or continued beyond it. For example, Meckel pictured evolutionary transitions between segmented worms and insects that paralleled the metamorphoses of maggots into flies, and between gill-breathing and lung-breathing vertebrates that paralleled the metamorphoses of tadpoles into salamanders.

Meckel did not believe in universal common descent, however. The diversity of life, he claimed, was "mostly original," the product of many separate spontaneous-generation events, at different times, places, and with different environmental conditions acting upon different mixes of substances. For example, the Australian marsupials were so similar to each other and so different from other mammals that they were likely to be the descendants of a different spontaneously generated organism. They probably had diversified and progressed to the mammalian level on their own separate evolutionary track. Fossil forms that had no living counterparts might also have descended from separate spontaneous generations or creations. But by the same token, *any* given form might have descended from a separate production and been unrelated genealogically to others that resembled it. There was no way to read ancestry from morphological resemblances, and on the whole, Meckel made little use of species transformation to explain *Mannichfaltigkeit*. He preferred to attribute more of the diversity of forms to originally generated differences.[68]

Thus, Meckel's recapitulationism was designed primarily to connect the embryo to the animal scale. The extension to the history of life or species transformation was much weaker and more tentative. His evolutionism was quite limited, and it was not supported with evidence from embryology. Meckel did not think that embryonic recapitulation revealed genealogical relationships, because similar patterns of addition, condensation, and differentiation were to be expected in unrelated groups, and because taxonomic groups could contain many unrelated evolutionary lineages.

Karl Ernst von Baer as Revolutionary and Reactionary

Meckel's fears of being misunderstood to have made nonsensical comparisons of whole embryos to lower adults or ancestral species, and of pathological forms to early embryos, proved to be well founded. Albeit without mentioning them by name, Karl Ernst von Baer ridiculed his predecessors for making just those comparisons, and it clear that Meckel and Oken are the intended targets. Ever since Baer's 1828 *Entwickelungsgeschichte der Thiere: Beobachtung und Reflexion* (Developmental history of the animals: observation and reflection), Meckel has been associated with simpleminded scales of nature, rigidly linear paths of development and evolution, and transcendental archetypes, while Baer has been credited for the concept of differentiation or specialization. It was a very successful attempt at intellectual patricide.

A landmark of descriptive embryology, Baer's *Entwickelungsgeschichte* began with a detailed account of the development of the chick, from the first hours after fertilization until hatching twenty days later. But after providing this unparalleled base of empirical evidence, Baer turned from "observation" to "reflection" and, in a series of "scholia," attempted to reform the fundamental principles of his science. Here we see him putting together a synthesis of developmental laws, transcendental archetypes, and the concept of orderly and purposeful *Entwicklung*.

Baer's embryology removed any doubts that might still have remained about the epigenetic production of form and the falsity of eighteenth-century preformationism, but it replaced the preexisting physical forms with transcendental ideas. He argued that the idea of the future adult had to be present in the embryo from the start to guide its progress. Blind laws and mechanistic processes, although surely important as intermediate links in the causal chain, could not do everything that was required. Baer clearly undid the efforts of Blumenbach, Kielmeyer, and Meckel to put transcendental ideas aside and to work on the observable patterns of change and give them physicalistic explanations. With Baer, ideas returned to the foreground as causal and directing agents.

Baer argued that mechanics alone could not describe development. Even if one had complete knowledge of the laws of development and the physical configuration of the embryo at any stage, one still could not predict its further development. The goal of an *Entwicklung* also had to be reckoned as its necessary cause: "Even though every step forward in development is only made possible by the preceding state of the embryo, the entire development nonetheless is governed and guided by the whole essence of the animal that is coming to be. The state of the embryo at every moment will not be the sole and absolute determinant of the future. If this, namely, is clear in and of itself, still it is not uninteresting to be able to prove this relationship through observation."[69] The embryo was not a complex machine as envisioned by Kielmeyer, in which a change in one part triggered a cascade of mechanistic changes throughout the system and made it develop. Such changes could not be counted on to proceed in a well-regulated fashion toward a consistent goal.

Although Baer took his teleology as axiomatic, he also offered the following empirical justification for the assumption. He observed greater variation in form among chick embryos than among adult chickens. Therefore, he reasoned, there must be something that keeps steering development toward the norm, making the variants converge on the proper adult form. In contrast, the mechanistic hypothesis would lead one to expect every variant embryo to go its own way:

From this, however, one can see that it is not each momentary state of the embryo, by itself and in all its details, that determines the future state, but rather that more general and higher relationships rule it. In this way, I believe, the natural sciences, which people like to accuse of favoring and nurturing materialistic views, can refute strict materialist teachings through observation, and prove that it is not matter and the way it happens to be configured, but rather the *essence* [*Wesenheit*] (or the idea, following the new school) *of the reproducing animal form that controls the development of the embryo.*[70] (Emphasis in original)

The direction of development, according to Baer, was not from species lower in the animal scale to higher. An animal was a member of its own type and species from the moment of conception and did not have to pass through the forms of lower species or types on its way to its adult form. Rather, the embryo became increasingly well differentiated, displaying more and more of the characteristic, specialized parts of its archetype. It reached a higher *Grad der Ausbildung* (Meckel's term for degree or grade of development) within the type, but it never changed types.[71]

The embryo could thus increase in complexity and express its essence or type in ever greater detail, but it could never express any other type.

What other authors might have described as recapitulation—for example, the appearance of a fishlike stage with gill arches and slits in an embryo destined to become a mammal—Baer explained as part of the process of differentiation. Fish and mammals followed approximately parallel developmental pathways, along which they expressed the common features of the vertebrate type to which they both belonged, such as the vertebral column and the gill slits. But the mammal was never identical to any particular fish or fish embryo.

Baer's technical arguments, wit, and sarcasm were a powerful combination that made recapitulation, along with any theory of evolution based upon it, seem quite ridiculous. With his droll humor, he skewered recapitulationists for the ease with which they let species move up and down a linear scale of nature like a railroad train on a track. However, in doing so he also reduced recapitulation to a caricature. All the subtlety and flexibility of Meckel's system, the analysis of differentiation at the organ-level, the variations in tempo and timing, and the provisions for diversity and variation, were lost. Baer was having some fun at the expense of his rivals and he admitted it freely. He even warned in a footnote against taking his own account seriously as a fair historical picture.[72]

Historians of biology from Russell on have ignored the warning and perpetuated the caricature as if it were an authoritative portrait. Gould and Bowler use it when they associate pre-Darwinian evolution and recapitulation with a linear scale of nature, teleology, and biological determinism. And although Richards connects Darwin to those same models, he still follows Baer's characterization of them, and he cites Baer's caricature uncritically as evidence for the emergence of evolutionary theory from early German recapitulationism.[73]

Baer asserted that the whole idea of species transformation was nothing but the reification of an ill-founded, continuous, linear scale of nature. Where once morphologists imagined the embryo going through an ideal sequence of forms, Baer said, transformationists now interpreted that same sequence as the history of the species. But who actually did this? Baer's straw man had no resemblance to Oken or to any other non-transformationist, since they never made such historical claims at all. The critique also did not apply to transformationists like Kielmeyer or Treviranus, who envisioned a discontinuous or branching family tree in place of the old scale of nature. Lamarck might have been accused of making the course of species transformation follow a linear scale of nature, but even his view allowed some branching, and in any case, he did not make the connection to embryology that Baer was trying to skewer.

Even Meckel, as we have seen, did not treat the scale of nature in the way that Baer insinuated. He did not insist on a strictly linear scale, and he did not have the embryo progress as a unit and recapitulate whole lower adults. Each organ or organ system progressed at its own rate, and its manner of progressing—Meckel's yardstick for measuring height on the scale of development—was not very different from Baer's at all. Using terminology usually associated with Baer, Meckel was already going by *Grad der Ausbildung* twenty years earlier. The main difference is that Meckel used his yardstick to make comparisons among any and all animals, whereas Baer allowed comparisons only within each of the four Cuvierian classes, which he also claimed to have described independently of Cuvier.[74]

In Baer's depiction, recapitulation theory stood or fell with the single straight-line progression of adult forms in the scale of nature. Baer objected that the animals could not be fitted onto a continuous linear scale, and he proposed a system of four classes, equivalent to Cuvier's *embranchements* only defined not by adult anatomy but rather by the earliest discernible symmetries of the embryo. None of the classes was higher or lower than the others, so there was no linear scale to recapitulate in the first place. The distinctness of even such early embryos of all the classes spoke for divergence from a common starting point, rather than recapitulation of a common line of development. Then there was the objection that some embryonic structures, such as the placenta, were never found in lower adults at all. So it should have been obvious that embryos were never perfectly equivalent to lower adults. Even worse, how absurd it was to think that an embryo could pass through the stages of air-breathing forms or flying birds or insects when it was immersed in the amniotic fluid![75]

But these were all problems that Meckel had dealt with. He did not have the embryo as a whole recapitulating, let alone "being," a lower adult. Only individual organs or organ systems recapitulated lower configurations, and each organ differentiated and advanced at its own rate, independent of the rest. When the ventricle of the heart was still incompletely partitioned, he could still describe it as being at the three-chambered reptilian level, even if there was nothing at all reptilian about the placenta. Besides, the placenta fit into Meckel's system quite nicely. He counted it as part of the respiratory system because of its role in oxygenating the blood, and it condensed and differentiated like any other respiratory organ. When the embryo lost it and switched to lung-breathing at birth, that paralleled the way in which many other embryonic adaptations were lost, as, for example, the external gills of the tadpole.[76]

From Embryology to Evolution

The main problem that Baer did create for recapitulationists of all stripes was his new theory of the direction of development, with its strong dichotomy between differentiation and linear progress, and its rejection of progress. According to Baer, an embryo differentiated from the general to the specific while staying within a single type. It exhibited more and more of the type's and its subgroups' characteristic features, but never recapitulated an evolutionary history or any other upward path through a series of types. Every vertebrate embryo, for example, was already essentially a vertebrate at the very start of its development and bore no trace of lower forms or ancestry. Consequently, there was no single developmental pathway that all species had to follow, with optional stopping points. Each species had its own exclusive path which it followed from beginning to end. It might run close to that of related species to some extent, but was never precisely parallel or overlapping, except perhaps at the very outset. Within a class, all embryos started out from a simple common point, but different subgroups soon followed different options for further differentiation. At each successive stage, there were more options that further separated groups within groups. Taken together, all the possible lines of development formed a branching pattern.

Baer sketched out a map of the branching pathways available within the vertebrate type. The beginning of the road was a generalized stage, common to all vertebrate embryos. Next came a multiway intersection, offering pathways to each of the vertebrate classes, such as mammals, birds, fish, and reptiles. All members of a class would take the same turn and develop the distinguishing common features of that class. From each class on the map, the pathways fanned out again into the various orders, and similarly to the families within the orders, genera within families, and so on. In the case of Baer's object of study, the chick, the embryo first took on the common features of its vertebrate type, then the common features of the class of birds, and so on down through the order to the family, species, and variety of chicken. However, from the earliest visible stage of development, the embryo contained the idea of the chick and already was a chick. It never had to rise through the ranks of other groups to which it did not belong.

There was no lower or higher in Baer's scheme of diverging developmental pathways, only grades of progressive differentiation within each pathway. Depending on how one drew the branching tree of development, one could place the endpoint of any desired branch at the top of the page and call it the pinnacle of creation. Baer could easily rotate his map to bring the birds to the top, in which case he joked that a recapitulationist bird might take the following view of mammals:

Those four- and two-legged animals have many similarities to embryos, for their skull bones are separate. They have no beaks, as we do in the first five or six days of incubation. . . . Not a single true feather sits on their body, but rather only thin feather-shafts, such that we are already further along in the nest than they ever come; their bones are only slightly brittle and contain, like ours in youth, no air at all. . . . A crop is entirely lacking. The proventriculus and gizzard merge more or less into one sack. These are all conditions that we pass through rapidly. . . . Their nails are so awkwardly wide, like ours before we hatch. The ability to fly is shared only by the bats, which seem to be the most perfect of them. And these mammals, who so long after birth still cannot seek their own food and never raise themselves freely from the surface of the earth, think they are more highly organized than we?[77]

It was an important challenge to embryologists, evolutionists, and systematists to be more precise in formulating their measures of progress and perfection.

Conclusion

Baer's story of a facile intellectual path from idealists' linear scales and recapitulation theories to species transformation was rather facile itself, and one must be wary of taking Baer as an authoritative historical source. The recapitulationists were not all transformationists, and the transformationists not all recapitulationists. Recapitulationists sometimes related embryonic development to an abstract scale of types, as in the case of Oken, but more usually they drew parallels to levels of complexity as defined by the acquisition of new faculties, degrees of differentiation and specialization of organs, or both. Recapitulationism was primarily about comparisons between embryonic stages and living taxa. When historical changes were held to parallel embryonic ones at all, as in the case of Kielmeyer, the parallels were not morphological.[78]

Historical parallels were more widely debated in the British context, following the publication of Robert Chambers's *Vestiges of the Natural History of Creation* in 1844, but even there, they were drawn between levels of complexity or, in the manner of Meckel, between individual organs, rather than between whole embryos and ancestral species. The parallels attested to the unity of the underlying laws and plan of creation, not to genealogy.[79] The book was not taken very seriously among German academics, and its translator, Carl Vogt, peppered the German edition with enough sarcastic footnotes to dampen popular enthusiasm for it.[80] In the 1850s the Swiss-American zoologist Louis Agassiz gained greater academic respect for a historical version of recapitulation, which put forward a threefold parallelism among the embryo, the animal scale, and a succession of fossil faunas. But here, again, the parallels expressed an underlying law or divine plan rather than family resemblances.[81]

Quite simply, no one thought that morphological similarities were good evidence for common ancestry, because there were too many other determinants of form besides history and heredity. Those biologists who came to believe in the transformation of species did so for a variety of reasons other than the mere suggestiveness of embryonic development.

Generally, biologists first had to become convinced of species transformation, based on other sorts of arguments, before they could even begin to see evidence of common ancestry in embryos. The intellectual path to evolution did not run through idealized types and recapitulation theory, but through comparative studies of anatomy, paleontology, and biogeography, adaptation, variation due to environmental effects, and the contemplation of *Mannigfaltigkeit*. Darwin based his case for evolution on these broader considerations, and pre-Darwinian German morphologists were making inroads in all of these areas as well. Pander and Meckel followed Lamarck in stressing environmental effects and adaptation. The Austrian paleobotanist Franz Unger is another good example of a late pre-Darwinian thinker, who brought in biogeography, reproduction and cell division, variation, adaptation to local conditions, and sequences of fossil floras to make a case for species transformation in the 1850s, into which he also incorporated embryological terminology and metaphors.[82]

Bronn, as the next chapter will show, was investigating a similarly broad range of phenomena to Unger's or Darwin's, relating to variation, adaptation, stratigraphy, and *Mannigfaltigkeit*. He rejected transcendental archetypes, linear systems of classification, and recapitulationism in favor of Cuvierian types and Meckel- and Baer-like criteria of differentiation and progress, and came to the conclusion that a full account of form had to look outside the organism and its internal structures and laws of development to include environmental change and adaptation. He was therefore more than open to new suggestions like Darwin's, as long as he could reconcile them with his standards of *wissenschaftlich* procedure and explanation.

Looking farther ahead, we will see that Haeckel took a similar path to evolutionism, reading Darwin (in Bronn's translation) just when he was grappling with the problem of variation and classification in the Radiolaria, which he happened to be studying in 1860. In his first published remarks about Darwinism, in his monograph on the Radiolaria, Haeckel seconded Darwin's arguments about variation in nature, and was unconcerned with the consequences of Darwinism for types and developmental sequences.

2
H. G. Bronn and the History of Nature

As late as the 1840s and 1850s, leading German biologists were still rebelling against dry description and classification—"hay collecting," Matthias Schleiden called it—and claiming higher intellectual goals and a higher status for their disciplines.[1] Schleiden expected his cell theory and its accompanying methodological prescriptions to uplift the practice of botany, and with it, human culture as a whole. In his pointedly titled textbook of *wissenschaftlich* botany, he predicted that soon, Man "as a thinking spirit will delve into the mass of phenomena, attempt to become aware of its inner lawful coherence, and so elevate himself to the level of *Wissenschaft.*" The idealistic and *naturphilosophisch* approaches of the earlier part of the century had not risen to this level, because they still tried to fit the mass of phenomena into preconceived categories, rather than practicing proper inductive methods.[2]

Bronn's Austrian colleague, the paleobotanist Franz Unger, was also concerned about advancing the position of his discipline among the natural sciences. Reflecting on the state of the field in 1852, he worried that botany might not keep pace with the other sciences, which were gaining rapidly in cultural importance: "It is not to be denied that the natural sciences currently are pushing themselves ahead of all the *Wissenschaften* in the arena of intellectual prowess and most of all are pointing out the way that the human mind is trying to take in its development." The key to progress in botany, he argued, was to continue moving to higher intellectual pursuits than the mere description and classification of plants: "The modest demands that were made of the plant expert have expanded significantly. Familiarity only with the coiffure, uniform, rank and position of the descendants of the most fertile goddess of Greece [presumably Chloris or Flora, the goddess of flowers], which in Linné's day still formed the keystone of all botanical erudition, now, after extensive rebuilding of the scientific superstructure, has been moved more to the side or used, in only a limited sense, as a basic foundation." All the extensive plant collections "could only be useful as material for a scientific investigation that has yet to be undertaken," and like most earlier biologists, Unger took the physical sciences as the model of how to proceed: "This fortunate turn of events in botany has occurred only recently, and the success of its efforts is assured insofar as it tends to lead to a *physics of the plant organism*" (emphasis in original).[3]

Bronn was in a comparable position, campaigning in similar terms on behalf of zoology and paleontology. His long road to becoming head of Heidelberg's first institute of zoology (only the fifth in Germany)[4] surely made him aware of his field's tenuous status as a mature *Wissenschaft*, worthy of a place among the university disciplines. In *Handbuch einer Geschichte der Natur* (Handbook of a history of nature) in the 1840s, he took pains to dispel the impression that paleontology was still in a prescientific stage of collecting and classifying. In 1850, his *Allgemeine Zoologie* (General zoology) aimed for a logical organization and synthesis of anatomy, physiology, zoochemistry, biogeography, paleontology, animal psychology, and traditional systematics into a comprehensive *Wissenschaft* of animal life, and his famous series on the *Classen und Ordnungen des Thierreiches* was motivated by a similar desire to coordinate and synthesize information of various sorts about each class and order. As the subtitle of the series indicates, the classes and orders were "depicted scientifically" (*wissenschaftlich dargestellt*).[5]

Bronn's approach to the history of life, as developed not only in *Geschichte der Natur* but also in the mature reformulation of his theories in 1858, adhered to the ideals of *Wissenschaft* that he had been developing throughout his career and incorporated many of the ideas reviewed here in chapter 1: the rule of law in biology; the unity of the biological and physical realms; the generation of *Mannigfaltigkeit* and historical change by the complex interactions among biological laws and forces and environmental factors; functional rather than transcendental types; and quantitative biogeography.

But there were innovations as well. Bronn was searching for multiple and preferably quantifiable measures of progress and perfection, independent of the embryologist's sequences, for use in an independent discipline of paleontology. As a paleontologist, he also had to deal with phenomena that did not occur in individual embryos in protected wombs or eggs, with stable environments. Explanations and laws modeled too closely after embryology failed him, as he tried to explain adaptation to changing environments, the extinction and replacement of forms over geological time, or the geographic distribution of past and present species and varieties. In 1859, he saw a potential ally in Darwin, who addressed many of his concerns and brought together a similarly wide range of perspectives on variation, adaptation, and change. But Darwin also undermined some of his assumptions about law and order, progress, and *wissenschaftlich* procedure, and threatened to make his successional system obsolete. This chapter will consider Bronn's intellectual commitments in detail, which will help explain his interpretation of Darwin's *Origin* and the issues he raised in response to it. First and foremost was the problem of how to approach the history of life in a properly scientific manner.

Putting *Wissenschaft* to Work in Paleontology

At the heart of *Geschichte der Natur* were hundreds of pages of tables, listing fossil species and counting and classifying them taxonomically, stratigraphically, and geographically. But the dry compilation of data was accompanied and justified by a theoretical and methodological framework that interpreted it and allowed Bronn to assert the *wissenschaftlich* nature of the entire enterprise. Bronn's framework drew on most of the principal elements of scientific biology discussed above, relying on special laws and forces as did Blumenbach, the systems thinking of Kielmeyer, the integration of biology with geology, chemistry, physics, and cosmology in the manner of Reil, and the quantification of patterns and trends in the fossil record, using the methods of Humboldtian biogeography.

There were also some conspicuous departures from earlier trends in morphology, however. The concept of archetype came in for criticism and was used only sparingly, and never in the sense of a transcendental idea or cause of development. When Bronn did write about types, as we shall see below, he referred either to actual organisms—the founding members of a species or larger group—or to the basic geometries or body plans of the major Cuvierian groups. Moreover, Bronn treated embryology quite separately from systematics and paleontology, and paleontology emerged as a *Wissenschaft* in its own right. It still exhibited the characteristics of an orderly and progressive *Entwicklung*, but it had its own independent trends and laws that had to be discovered in the fossil record and not in comparative embryology.

Bronn positioned himself as a *Wissenschaftler* at the outset, explaining what a properly scientific approach to natural history entailed. As a teacher of applied natural history and cameral studies, he did not insist on the purity of research and its detachment from practical affairs, but he did require a strong theoretical component. This component needed to be broad in scope, in Bronn's opinion, to synthesize and illuminate large collections of empirical information, and to transcend the mere collection and classification of data and compilation of case studies. It could not be satisfied with ad hoc explanations of individual cases, but had to seek general laws. As a first approximation, a law could be an observed trend or an empirical rule of thumb, but a proper *wissenschaftlich* analysis moved from such rules toward the establishment of deterministic or logically necessary relationships.

Bronn began his own history of nature by establishing the comprehensive scope and unifying principles of his enterprise. Nature was a system of complex interactions. One could not explain biology without geology, or geology without astronomy and cosmology, he argued. That was because the Earth was a planet, and so its geological history had to begin with its formation out of a cosmic nebula. The subsequent history of the planet, as explained by physics, geology, and meteorology, accounted for the conditions that supported life and to which life continually had to adapt. Biology, geology, and cosmology were therefore all aspects of a complex, dynamic whole. The book underscored that "in space and time, organic life depends upon inorganic, telluric upon cosmic, too necessarily and in too many ways to allow a *wissenschaftlich* treatment of geology nowadays to do without astronomy and a history of plant- and animal creation to do without geology; and that one can find the preconditions for the phenomena of the one for the most part only in the other,—that consequently it will have been essential to take astronomy as the starting point for the history of nature."[6]

Bronn further unified his view of nature by inferring the existence of a single hierarchy of laws and forces. Like Reil's, his hierarchy featured physical, chemical, and biological forces that acted in increasingly complex and selective ways. But where Reil wrote generally about how the different mixtures and arrangements of materials and their associated forces could account for organic form and development, Bronn tried to be more specific about how that might have worked. For example, the gravitational force acted at any distance upon any and all materials and tended to pull them as closely together as possible. Geometry required that spherical shapes be produced by such symmetric pulling, which accounted for the shapes of raindrops and planets and various other globular and concentric forms. Not only that, but the very fact that gravitation had been able to pull the planets into spherical form was proof of their originally liquid state.[7]

The forces of chemical affinity or elective attraction acted only over short distances and only between special qualitative and quantitative combinations of materials. They did not merely strive to pull particles nearer together in space, but to unite them into homogeneous compounds. They were responsible for the forms of mineral crystals and for prismoid shapes in general. Many different compounds and specific shapes could be produced by the actions of the chemical forces, and even more by the vital forces. These generated the many and diverse forms of plants (which are all variations on a basically conoid theme), and animals (basically sphenoid, or wedge-shaped).[8]

Thus, for Bronn, the science of life was built upon physicalistic foundations. It began with the geometric or architectural problem of how to generate complex forms from the more basic shapes and physical elements that were made available by the actions of fundamental forces. Later he discussed additional determinants of form, such as the need for functionality and adaptation to a given environment, and the laws governing progress and the production of variety. As with Kielmeyer and Cuvier, there was no talk of transcendental archetypes or ideas that guided development, but only of forms being determined by complex mechanistic interactions and practical requirements.

Another of Bronn's methodological concerns was with the role of descriptive natural history in a *wissenschaftlich* treatise. The bulk of Bronn's *Geschichte der Natur* was taken up with tabulation of all the known fossil species and where they had been found. That alone made it a valuable reference work, but Bronn was not content with description and classification, and he took pains not to let the presentation of these data become an end in itself. In his role as *Wissenschaftler*, Bronn promised "systematic organization and scientific illumination [*systematisches Ordnen und wissenschaftliche Beleuchtung*] of purely factual observations."[9] Systematic organization of the facts would reveal patterns of historical change, and, under proper scientific illumination, those patterns would yield the laws of change.

As he organized and illuminated his data, Bronn professed a purely inductive ideal of scientific reasoning and said he was making the first attempt ever to develop a comprehensive history of nature directly from the facts, without any preconceived theory. One may well doubt his ability to banish all preconceptions from his work, but he did succeed in identifying and demolishing particular ones he thought had been misleading in paleontology. For example, Bronn criticized other authorities for letting their theoretical precommitments influence the way they classified fossil specimens. His main targets were catastrophists like Louis Agassiz, who held that cataclysmic events—floods, volcanism, or glaciation, but of a magnitude unseen today—wiped out whole floras and faunas at a stroke and created abrupt transitions between geological periods. Bronn accused them of improperly assigning different species names to similar specimens, just because they were found stratigraphically above and below a purported catastrophe. The classification then accentuated or even created the impression of widespread and sudden replacement of species.

For example, Bronn made an impressive empirical case for continuity of most species' life spans across the boundaries of geological periods, using data from every major animal group. Only the fossil fish seemed to conform to Agassiz's contrary expectations, but virtually all of the fossil fish had been classified by Agassiz himself, with his catastrophism in mind. By invoking the *wissenschaftlich* ideal, Bronn undermined Agassiz's authority.

Bronn was prepared to concede that present-day species were scarce in the Pliocene, as Agassiz's interpretation required, but he blamed that on a combination of the incompleteness of the fossil record and Agassiz's biases: "Only the lack of still-living fish species in Pliocene strata might seem surprising at first; however, one must remember that the youngest marine formations are not conducive to the preservation of fossil fish, but let them fall apart into individual vertebrae and bones. And besides, Agassiz, whom, so far, we have to thank for the identification of almost all fossil fish, seems in several cases at least to have differentiated the youngest (Pliocene) fossil fish species from living species by means of some very unimportant characteristics."[10]

Bronn also quoted Agassiz to the effect that he did not think morphological characteristics alone demarcated species, but that their dependencies on particular environments were more important. Bronn agreed up to a point, but his own system did not assume that species would always be perfectly adapted to their environments. On the contrary, he expected them gradually to become maladapted as the Earth evolved, and possibly to live on into later geological periods despite the (for them) deteriorating environmental conditions. Bronn thought that Agassiz assumed too close a correlation between species and environment and might have been tempted to say, as he compared two similar specimens, "Same stratum, same species; different stratum, different species." That would have been biased, even circular, since strata were often identified by the species they contained. It was better to go by morphology: "Under such circumstances, the most impartial way to proceed—not in conformity with a preconceived opinion, but leading toward a final result—would be to say: whatever one has no means for separating, is to be united in one species."[11]

Noncatastrophists might well have had a bias opposite to Agassiz's, toward lumping dissimilar specimens together under a single species name and thus obliterating any evidence of sudden, synchronous turnover. In compiling his own tables, Bronn frequently had to adjudicate between conflicting classifications of the same fossils by different authors, and it was important for him to set out his criteria for lumping or splitting species without reference to presumed patterns of change. Bronn insisted that he could and did decide strictly on the basis of morphology, without letting stratigraphic sequence influence him.

The Principle of Adaptation and the External Causes of Change

A corollary of Bronn's extreme inductivist stance was that he would not feign hypothetical causes of change. Bronn followed the example of Newton and the Göttingen school by looking for regular, quantifiable patterns of change and describing them in terms of laws and forces. He made no commitment about the nature of those forces or the reasons why they obeyed the laws. Most of the time, he claimed no more for his "laws" than that they were generalizations abstracted from repeating patterns in the data and no more for his biological forces than that they were the otherwise unknown causes of those patterns: "The 'laws' that we set up here are mere abstractions out of the sum of the observations to date—often only imperfect induction—without mathematical necessity. New observations can modify or overturn them."[12] The various laws of morphological progress in the fossil record that Bronn identified were clearly in this category of mere empirical generalization.

There was, however, a significant exception, and that was the "principle" or "law" or even "fundamental law" (*Grundgesetz*) of adaptation, to which Bronn ascribed a higher status. It was not an empirical rule, subject to reconsideration as more data were collected. It held *necessarily*, for both natural and a logical reasons, because it was evident that no species would survive if it appeared at a time and place for which it was not suited. At times Bronn seemed to imply that nature would not even create such a species in the first place: "The fundamental law, whereby the gradual appearance of the living world was guided has been that of the continual adaptation of the living world to the external, geological conditions of existence at every point in time."[13]

Bronn even suggested (inconsistently here, but more strongly in 1858) that the other laws of biological change might be redundant. They could perhaps be subsumed under the effects of progressive changes in the physical and biological environment, combined with the more fundamental law of adaptation that required biology to keep pace with geology: "Yes, we do not doubt that if we knew all the former conditions of nature precisely, this principle [i.e., of adaptation to external conditions] would still be the sole basis of all the phenomena, just as the phenomenon of gradual increase in perfection can already be subordinated to it."[14]

This was a radical break from earlier German ideas about morphological change, because it gave environment and adaptation priority over archetypes, internally generated *Bildungstriebe*, and other developmental processes that ran predetermined courses without reference to the external world. Bronn invoked no transcendental ideas or types that guided development or provided it with its goal.

With this necessary and incontrovertible principle of adaptation as his platform, Bronn was also in a position to attack alternative theories of gradual species succession for their arbitrary and implausible preconceptions. The ones he criticized were based on an idea associated with the morphologist and physiologist Johannes Müller and the paleontologist Giovanni Battista Brocchi, that species came into existence at various points in the fossil record, then went extinct after living out their individually predetermined life spans or using up their finite endowments of life force. Bronn counted the successional theories of Richard Owen, Charles Lyell, and Christian von Meyer as variations on this idea.

Bronn found them all implausible for their overreliance on blind, internal processes, within the organisms themselves, that were unresponsive to the external world yet just happened to yield adaptations when and where they were needed. Neither did it make sense to Bronn to give a species a predetermined life span, when its survival was contingent upon the availability of the environmental conditions to which it was adapted. Species depended upon one another in complex ways, for example for food or shelter, yet the life-span theories did not take such relationships into consideration. Unless one was willing to rely on divine providence to synchronize the life spans of interdependent species with each other and with the evolving Earth, there was nothing in these theories to prevent the whole system of nature from becoming unbalanced and going to ruin.[15]

Similarly, in Bronn's view, any models of organic history derived from embryology were also doomed to failure because they looked inward for the causes of change. They either discounted the need for adaptation to the changing external environment, or again left it to divine providence or the inherent harmony of nature to keep the organic world in synchrony with physical Earth. Agassiz's recapitulational system came in for special criticism on this score, but the same would have applied to most other pre-Darwinian German theories. Agassiz held that the progressive sequence of forms observed in the individual animal embryo paralleled or recapitulated the successive creations of forms within the animal's class, as recorded in the fossil record. Bronn did not dispute that both sequences were progressive—indeed he himself was out to prove the progressive nature of organic history—but they progressed along different paths, with different needs for adaptation. He saw no *necessity* for such a close connection between them as Agassiz wanted to make, and no reason to read them as consequences of the same divine plan of creation.[16]

Because they were not necessary, Agassiz's rules, if they applied at all, were of a different sort from Bronn's principle of adaptation and had a lower status. As Bronn argued: "Now that the phenomenon, referred to at the outset, of the gradual and ever-more-frequent appearance of more-perfect organic forms, in addition to those that had been originally present, has been verified, we inquire as to the necessity of this phenomenon, but without being able to find any such thing, neither in external causes, nor in the principle itself."[17]

In addition to not obeying the same kind of laws, the progress recorded in the fossil record also ran in a different direction from embryonic development: "For the young individual of an already-created species, setting out from its initially very imperfect condition, surely must run through certain increasingly perfect levels of organization, following the laws originally impressed upon the species, in order finally to arrive at the highest level possible for that individual."[18]

In short, species succession was quite distinct from embryonic development, according to Bronn. The embryo came into existence already belonging to a species, subject to species-specific laws, and with a predefined adult form to reach. New species, on the other hand, arose from separate creative acts, which were not dictated by any law or bound by any requirement to recapitulate embryonic forms: "In the case of creation, which appears to have been the result of a new act of the Almighty and not the result of a pre-existing law of nature, no gradual progression was required to go from an embryonic form of each class to that class's highest level."[19]

And even if there were a need to produce forms equivalent to all the embryonic stages, there was no reason for nature or the Creator to run through them sequentially, rather than create them all at once.[20] Bronn came to the conclusion that the causes of progressive development in the fossil record would not be found in embryology, and that they would not have the status of necessary laws. An element of chance and historical contingency had to be involved.

Accounting for Progress

Bronn's belief in fixed and necessary laws, along with his case against catastrophism, made him very sympathetic to the uniformitarian school in geology. He took for granted that laws and forces stayed the same throughout time, and that knowledge of present-day processes could and should be used to explain the past. However, Bronn rejected Charles Lyell's further stipulation that the Earth as a whole underwent no net historical change. Lyell's steady-state hypothesis held, for example, that erosion in one place would always be balanced by deposition in another, mountain-building in one place by subsidence of the crust someplace else. Change was thus local and cyclical, rather than global and directional. Applied to paleontology, steady-state assumptions led Lyell to a theory of species succession, in which repeated extinctions and creations yielded only directionless turnover, never any net change in the overall character of the fauna and flora, no global progress in complexity of organization. In opposition to Lyell, Bronn argued that such an interpretation was incompatible with cosmology and the physics of the cooling Earth, as well as with the unity of the sciences, as proclaimed at the beginning of the book. This is where Bronn's efforts at unifying geology with biology and cosmology came to fruition. They let him adjudicate between uniformitarianism and catastrophism, taking something from each school: progress from the catastrophists and uniformity of the laws of nature over geological time from the uniformitarians.[21]

Bronn devoted most of volume 1 to detailing his unifying view of geology, cosmology, and physics, by giving an account of how the Earth originated as a molten ball and subsequently steadily cooled. The cooling was what gave geological history a direction. Once it progressed far enough, a crust formed, thickened, shrank, and buckled. It broke open in places and allowed volcanoes to burst forth and further change the surface of the planet and even the composition of the atmosphere, by releasing minerals and gases. Later, water in the primitive atmosphere condensed, fell as rain, and filled lake and ocean basins. The actions of wind and water further differentiated parts of the originally uniform planetary sphere.[22] Only when conditions progressed far enough, Bronn explained in volume 2, could plants and animals appear.

"Only after the temperature of the earth's surface sank to a certain level, and air and sea took on a certain mixture," Bronn argued, "was the earth suitable to become a dwelling place for plants and animals."[23] The organisms themselves set further changes in motion, driven by the dynamics of a complex system of interacting parts and forces reminiscent of Kielmeyer's account of development. By producing humus and soils, for example, or by changing the balance of oxygen and carbon dioxide in the atmosphere, organisms altered the environment and created conditions required by other organisms. Species that were adapted to biotically generated conditions were therefore dependent on the activities of preexisting species, and—according to the principle of adaptation—could appear only at certain points in history. When they did appear, they in turn modified the world in some small way and helped cause a cascade of floral and faunal changes. The system could never settle into a Lyellian steady state.

Rather than address the "mystery of mysteries," as the British were calling it, of what caused new species to appear in the fossil record, Bronn left the matter more or less open. Bronn's first task as a *Wissenschaftler* was not necessarily to identify the cause, but to ask what empirical laws it might follow. There were quite a few laws of progress, adaptation, and diversification to be discerned in his stratigraphic data. The main thing, according to Bronn, was that the fauna as a whole gradually came to include more and more modern, complex, and "perfect" forms, and Bronn analyzed his concept of perfection into several more or less quantifiable components, and derived various "laws" from the patterns of progress he observed.

In the 1840s, in *Geschichte der Natur*, Bronn presented some morphological measures of progress, but he put much more work into documenting progress in taxonomic terms, by the changing composition of the fauna and flora. Using the methods of Humboldtian biogeography, he counted the classes, orders, families, genera, and species living in each of his six major geological time periods. The numbers fluctuated, but on the whole Bronn found, among other things, a general increase in the absolute number of groups and their subgroups. The ratios of species per genus also tended to rise over geological time. Aberrant numbers from certain geological periods were explained away as the result of random processes, unequal durations of the geological periods, and unequal suitability of conditions for preserving fossils. All together these data indicated a lawlike trend toward diversification.[24]

Bronn also provided an analysis of the complexity of organic forms and was able to demonstrate a form of morphological progress. He found that the higher and more complex forms within each major group tended to make their first appearances in the later geological periods: "The gradual change in the forms of the organic world is mediated by the gradual appearance of newer and newer form-types and by the gradual disappearance of a portion thereof. The organic remains found in the oldest strata, although they are already very diverse, nevertheless do not yet contain certain higher organizational types, which rather have come to light in the strata of later periods or even just in the current period."[25] Morphological progress occurred without the lower forms necessarily being eliminated or even becoming less numerous.

Note also that Bronn had no place in his system for a linear scale of nature. His progressive morphological trends ran separately within every major group. Plants and animals were equally old and became complex over time in independent ways. The Cuvierian classes had been four in number at all times since the earliest documented fauna, in the Silurian Period, and Bronn did not place any one of them ahead of, or above, the others. Moreover, for geological and ecological reasons, he did not think it could ever be possible for the fauna to develop straight up a linear scale of nature, even if there was one:

The reciprocal dependencies of the lower upon the higher and the higher upon the lower forms, and the geological changes in the external conditions of life of different classes would not even be the only obstacles to get in the way of such a simple developmental pathway [*Entwicklungs-Gang*]. It is in any case impossible for the systematist to arrange all classes of organisms onto a single scale. . . . The way of life [*Lebens-Weise*] of the organisms, their behavior with respect to the external world, certainly often has, however, exercised a more powerful influence on the ordering of its eventual appearance and extinction than its level of organization would have been able to do, even if the latter were the underlying, guiding principle.[26]

One special measure of progress that Bronn used was ecological and followed trends toward colonization of new kinds of habitats. Bronn showed that in most taxonomic groups the first sea creatures antedated the first freshwater creatures, which antedated the first land-dwellers and fliers. The higher taxonomic groups and better-developed faunas and floras were characterized by increasing proportions of freshwater and especially terrestrial species. This pattern was to be expected from the evolution of the Earth, the progressive availability of habitable and diverse environments, and the necessity for adaptation. In his later work he referred to it as "terripetal" development.[27]

Variation Out of Uniformity

Bronn's assumptions about law and order in nature were continually challenged by his experience of individuality and variation—*Manchfaltigkeit*, as he called it (which I have been translating as "diversity"). How could a small set of deterministic laws, acting upon an initially uniform, spherical planet, produce so many and varied things? Some *Naturphilosophen* and idealistic morphologists might have tried to look past individual, accidental variation to discover natural reality in abstract forms and types, rather than in actual specimens. To them, individual variation was an obstacle to clear understanding, rather than an object of study. For his part, Darwin took variation more or less as given and sought to explain how an orderly, purposeful organic world could be made out of this original chaos. In contrast to both approaches, Bronn treated variation as a striking fact that needed to be explained.

Bronn's solution was to derive diversity from complexity: "The organic force seems on the one hand to have a much more limited and superficial range and effect than the other, inorganic forces. And yet, its behavior in itself and in relation to the others is not inconsiderable, because, through its infinite individualization, it finds so countless many and manifold points of attack, brings forth so many crisscrossing conflicts." In other words, although Bronn's laws were few, they interacted with each other and with the Earth in complex ways that produced variety at all taxonomic levels, down to the individual. Bronn was fascinated with this important problem: "The further development of the oppositions of this force in itself and toward those of the other forces, the derivation of the unlimited diversity of the phenomena from the few basic forces of nature, is one of the most pleasant contemplative exercises."[28]

Ultimately, the Earth's evolving crust was the engine of variation and change. It set the dynamic, complex system into motion: "The more varied these conditions of existence became, in consequence of progressive differentiation of the seas, lands, and climates, the more varied the living world became as well."[29] Local conditions then influenced the way morphological laws, forces, and trends were expressed in the production of species and individuals. The law of adaptation to prevailing conditions always had to be satisfied, too, and it worked at cross-purposes with the other laws of progress to make each new species a unique compromise. Thus, Bronn allowed a great deal of unpredictability and historical contingency to emerge within his ostensibly deterministic system.

In addition to rejecting developmental schemes that constrained natural progressions to stay on idealized pathways and failed to do justice to diversity, Bronn also had no use for transcendental archetypes abstracted from present-day forms. He preferred Cuvierian functional types, but expressed some skepticism about them as well. He thought they, too, might have tempted the systematist to force fossil organisms into present-day categories, when in fact they were sometimes intermediate between, or combined features of, two or more. Even Owen and Agassiz, as Bronn pointed out, had observed this phenomenon and provided for a certain amount of differentiation of subtypes.

But the problem was much worse than the mere modification of archetypes over geological time. Archetypes did not even provide consistent standards of comparison at any given time, either. The lowest and oldest echinoderms, Bronn pointed out, were the crinoids, or sea lilies, and these were generally taken to represent the archetype of the whole class and its lowest level of complexity. But crinoids were many and varied, and some of them were quite complex in structure or had advanced features (such as an anus) that were lacking in the supposedly higher sea stars. The same held in other animal classes, where the more primitive groups, which were closer to the supposed archetype, might have exhibited unexpected and unappreciated complexity. Judgments about what was more or less complex were therefore best made by analysis of many individual features, rather than by taking any one group as archetypal and primitive.

On the whole, Bronn found the archetype concept to be of limited applicability, and because it was not universal, adherence to type did not rise to the status of a law, let alone a fundamental law. Bronn said the law of adaptation was empirically better founded, universally and necessarily obeyed, and had the power to bring forth species that broke the constraints of the type.[30] Thus even though Bronn was not an evolutionist, he allowed his taxonomic categories to have fluid boundaries, and he cannot be classified as an idealist or essentialist. Only in his treatment of species did Bronn try to maintain a stable, natural category, but even at the species level, there was no essence or ideal type.

In principle, Bronn would have liked to define a species by community of descent from one or more founding individuals or breeding pairs. Those founders did not correspond to any transcendental archetype or idea, but were themselves the archetypes or "prototypes." In practice, unfortunately, one could never know precisely how the founders looked or whether they were all exactly alike. Bronn argued that there would almost always have had to be many founding individuals or pairs, otherwise the species might not have been able to get established, especially if it was slow-breeding or subject to depredation.

The founding individuals, Bronn assumed, need not have been identical to each other: "No law can be discovered or imagined that would determine how much or how little two original individuals must resemble each other in order for them to have to be counted as members of a single species or different ones." Therefore, there was likely to be original variation within every species, and no single ideal form: "The concept 'species' loses the rigid fixity that it would otherwise have."[31]

In practice, one had to define a species by its range of morphological variation. One first had to establish the range by observing how much variation there was among individuals known to be of common descent. Then one could include within the species any and all other individuals that also fell within that range. Bronn made special allowances for geographic varieties, which arose when a species' founders were created far apart and with original differences. The test of whether varieties belonged to the same species was whether they could interbreed. When the test was impossible, as of course it always was in paleontology, one had to reason by analogy to established present-day cases and estimate how much variation the species definition could accommodate.[32]

In a section that was heavily annotated by Darwin,[33] Bronn provided an extended analysis of intraspecific variation, with special reference to the experiences of practical breeders in England. Bronn's approach differed from Darwin's in two important ways. First, he concerned himself much more with discovering the causes of increased variation under domestication than with analyzing breeders' methods of working with this variation once it occurred. This was perhaps the point of view of the cameralist and civil servant, whose role was to evaluate the breeders' work and provide the conditions and materials that were most conducive to it. For example, he counted the varieties of dahlias produced since the arrival of the first specimens from Mexico in 1790, calculated rates of variety-production, and looked for environmental factors that caused variation and might be capable of increasing the breeders' future productivity.

In contrast, Charles Darwin looked at variation with the gaze of the landed gentleman, who had not only an economic interest in scientific agriculture, but personal experience with the breeding of hunting dogs, horses, fancy pigeons, and even dahlias for pleasure. As Adrian Desmond and James Moore put it in their biography of Darwin: "He had grown up in the agricultural heart of England, where horticulture and husbandry magazines lay around every manor house. His mother had kept pigeons; Uncle Jos [Josiah Wedgwood, Jr.] was a leading sheep breeder who had introduced short-haired Spanish merinos into his flock. Uncle John Wedgwood cultivated dahlias and advised him on plant crosses. Landowners had a wealth of knowledge about producing domestic breeds to order."[34] Compared to Bronn, Darwin inquired more thoroughly into the breeders' methods and individual achievements, and was more deeply impressed by the transforming power of selection, from which he argued by analogy to the power of natural selection to create new species.

Stabilizing Species, and Generating New Ones

In further contrast to Darwin, Bronn focused on the limits to the breeders' art, rather than their history of accomplishment. Bronn observed their inability to create new species or even permanent varieties, and by analogy he set comparable limits to intraspecific variation in nature. To be sure, Bronn observed that breeders' varieties did sometimes break out of the range of the species' natural variation, but he located the causes of these extreme variations in the process of domestication itself, rather than in selection. The change from the natural environment could destabilize the species, and artificial crossing could introduce even more new variation. He also said that such anomalous variations could occur in nature as well, but less frequently. At first blush, Bronn acknowledged, the evidence from artificial breeding might have seemed to suggest that there was no limit to variation and no fixed boundaries between species: "As one can see from the previous sections, nature has hundreds of means, some external, some random and unknown, of calling forth new variations [*Abänderungen*] from existing species. Thus, if one wishes to consider the production of monstrosities as well, there is almost no kind and no degree of transformation [*Umgestaltung*] that would seem in and of itself to be impossible."[35]

Still, nature was orderly, and Bronn would not let such variation result in taxonomic chaos. In the long run, he saw countervailing processes returning the species to its normal range of variation: "On the other hand, nature possesses a perhaps more limited number of means, but less random and much more lawful ones, of hindering the original species-type from going under among new forms."[36]

These "lawful means" of returning species to normal were the basis of Bronn's argument against species transformation. He saw environmentally induced changes as temporary. When the environmental factor ceased to act, the changes disappeared from the species, just as the work of the breeder came undone when a domestic variety was returned to the wild. Generally, anomalous or monstrous individuals would not survive long enough to reproduce, because of the imperfection of vital organs. Or if they did survive, they generally would have only more typical individuals to mate with, and so would have trouble passing on their anomalous variation. And even if the offspring remained atypical, there would still be a large number of hurdles to their survival and reproduction, including sterility and lack of vigor. Inborn variations "mate just as gladly and successfully with the typical form as among themselves, and they transmit [*übertragen*] their peculiarity to their offspring quite often, in the former case—where it must struggle against the more powerful, because more stable, type,—but in the latter case almost always. However, many of the forms that arise in this way must soon perish, because they are based upon imperfection of organs that are essential to life."[37]

Bronn's claim to have anticipated Darwin's theory[38] could very well have been based on this process, which amounted to a limited form of natural selection. Bronn conceived the natural world as a dangerous place, where much loss of life was required to maintain the balance of nature in the face of Malthusian overproduction. He even characterized the interrelationships of species with each other and with inanimate nature using words such as *Konflikt* or *Kampf*, a wording that foreshadowed Darwinian "struggle," which Bronn would later translate as *Kampf*. He wrote, for example, that seeds had to struggle (*kämpfen*) against dry summers or severe winters.[39]

In contrast to the Darwinian version, however, the outcome of Bronnian struggle was not gradual, progressive change, but only the demise of extreme anomalies, maladapted hybrids, and, in geological time, obsolescent species. It was a conservative and eliminative process, not a creative one. Bronn did not apply the idea to the differential survival of all the minor variations within a species.

Like species, whole faunas also had some capacity to maintain stable characteristics over the long run. Following minor disturbances, they, too, returned to normal: "Thus every region receives not only its fitting fauna and flora, but within them also a certain locally appropriate balance of species and individuals, which, if ever it is disturbed by random fluctuations of the causal forces [i.e., environmental influences], will always return by itself."[40]

But here the stabilizing process was not as lawlike as it was at the species level. Some geological changes, especially the progressive kind, could upset this equilibrium and set lasting, directional changes into motion, involving extinctions and creations of species: "Brute force cannot change it [i.e., the regional balance] in any individual respect without these changes then bringing on further changes in the rest and in the whole. Such unusual and violent moments could be: continental uplift, changes resulting therefrom in air- and sea-currents, or from the constructions of corals and so forth, so that in this connection, too, the earth is not to be viewed as being at rest."[41]

The result was a dynamic but orderly and coordinated development of the physical and biological worlds. Left on its own, the biological tended toward an equilibrium in the composition of the flora and fauna and the relative abundances of species. As significant geological change accumulated, however, the system lost stability and underwent directional change. Species that were no longer well adapted began to go extinct. Other species that were dependent upon them soon followed them into oblivion and created openings for new and better-adapted species.

The Production of Species

Where do the new species come from? Bronn did not like the idea that they evolved from preexisting species. He discussed just one evolutionary alternative to his successional scheme, namely Lamarck's, and he was very critical of it. First of all, he objected to Lamarck's decoupling of biological from geological evolution, for if evolutionary progression occurred at all, its causes would have to be sought in the environment to which life had to adapt. For Bronn there was no physical or logical reason for biological change to occur on a static Earth, and he did not see geological evolution reflected in Lamarck's system. He also demanded empirical support for Lamarck's claims that primitive organisms were produced continually by spontaneous generation, that environmental forces and "subtle fluids" had the power to cause lasting modifications in a species, or that there was a steady progression of species up a linear scale of nature.[42]

In short, Lamarck failed to live up to Bronn's high principles of *Wissenschaft*. Lamarck invoked preconceived, hypothetical mechanisms, described patterns of historical change that did not emerge from a systematic collection of stratigraphic data, and formulated laws of change that were neither generalized from empirical trends nor "necessary" like Bronn's law of adaptation. By Bronn's reading, Lamarckian evolution actually violated the law of adaptation, because it let the same primitive organisms be spontaneously generated at any time and place and evolve up the same scale of perfection without regard for the prevailing environmental conditions.

One might have objected that Bronn was vulnerable to some of the same criticisms, such as lack of empirical support for any mechanism he could think of for making new species come into existence, but Bronn did not acknowledge this as a problem. He was reserving judgment, but evidently thought that whatever the mechanism eventually turned out to be, it would be very different from spontaneous generation as envisioned by Lamarck. Bronn's mechanism had to generate every species, not just the most primitive ones, and it had to be sensitive to environmental conditions and capable of performing differently at different times and places.[43]

Bronn's position on the origin of new species shifted over the course of his career, but in *Geschichte der Natur*, Bronn still favored divine intervention as the cause of species production: "We recognize . . . in this appearance [i.e., of the first organisms], in the way in which all the simultaneously existing beings and all those that gradually follow one another are bound together, as well as in the wonderful organization of so many and varied organisms and in their adaptation to the external conditions of life at every time, a realized idea, such a planned process, such an appropriate interlocking of all the interacting conditions, that these things all together and individually can only be the effects of an infinite omnipotence and the arrangement of an inconceivable wisdom."[44] Elsewhere in *Geschichte der Natur*, Bronn said that God's will was the source of "the whole of objective Nature, with all the laws that lead it onward on a consistent course."[45]

Ideally, Bronn would have liked his divine Lawgiver to operate in a manner consistent enough to be treated scientifically like a force of nature. Bronn sometimes referred to Him as the *Urkraft* (ultimate or primeval force) or used ambiguous formulations such as *Schöpfungs-Kraft* (creative force), apparently in an attempt to sound more like a natural scientist than a natural theologian. Bronn was not out to prove the existence of God and infer His attributes from His decisions about creation. Rather, Bronn's investigations began from the assumption that God had already organized nature, endowed it with laws and forces, and that these needed to be discovered and described.

Even though Bronn's God had to intervene continually to keep the Earth populated with well-adapted species, His actions were not arbitrary. They could be studied in a scientific manner because He followed a plan and endowed nature with law and necessity: "No chance coming into being and no generation [of new species] has occurred, but rather original, intentional, planned creation by an absolutely independent Creator, unlimited in time, omnipotence and wisdom, Who places in all the Creation the necessity [*Nothwendigkeit*] of individual demise and at the same time the capacity [*Fähigkeit*] for rejuvenated reappearance in a different individuality, hence for continual existence amid eternal change and passing away."[46]

Note, however, that Bronn had only *demise* as a natural and logical necessity. When changes in the evolving Earth made a species cease to be well adapted, the fundamental law of adaptation went into effect and the maladapted had to go extinct. Thus the eliminative side of the historical process was brought fully under the control of well-founded and neces- sary laws. The same could not be said of the creative side. Nature had the *capacity* to bring forth new species, but Bronn could discern no necessity for her to do so. He could only say that *when* new species appeared, they had to be well adapted to prevailing conditions, and their addition would tend to advance the fauna or flora on his various scales of progress. There was still some room for chance or divine willfulness to initiate the crea- tive process and form the new species.

Bronn's Essay of 1858

Bronn completed the third and last volume of *Geschichte der Natur* in 1849, and then turned to other projects, such as his *Allgemeine Zoologie* and the inaugural volumes of *Classen und Ordnungen des Thier-Reichs*. When the French Academy announced its essay prize competition in 1850, Bronn did not respond, even though the questions the Academy posed were perfectly tailored to his research interests and his approach in *Geschichte der Natur*. They called for a study of the laws by which fossil forms were distributed through geological time, the timing of the appear- ance and disappearance of species, the relationships between species' life spans and the boundaries of geological strata, and whether new species arose by transformation or creation.[47] Only four manuscripts were sub- mitted at the 1853 deadline, and none was deemed comprehensive enough for the prize. The call for essays was reissued and the deadline extended to the beginning of 1856, and this time Bronn responded. Bronn's essay, completed at the end of 1855, was awarded the prize in 1857 and was revised, translated, and published in 1858 as *Untersuchungen über die Entwickelungs-Gesetze der organischen Welt während der Bildungs- Zeit unserer Erd-Oberfläche* (Investigations into the developmental laws of the organic world during the formative period of our earth's surface [hereinafter *Entwickelungs-Gesetze*]).

In *Entwickelungs-Gesetze* Bronn made much the same case as before in favor of gradual change and progress, and the same arguments for species succession and against transformation, but there were some significant changes as well. He now analyzed the concept of progress more thoroughly and provided new measures of it. He simplified and codified his array of developmental laws. And he changed his mind about the admissibility of the Creator in a *wissenschaftlich* explanation of life, opting for a naturalistic "creative force" instead.

Progress, as measured by the changing species composition of whole faunas and floras, was no longer the focus of the presentation. Bronn had been concerning himself much more with anatomy and comparative morphology than in the 1840s, and he had improved his tools for comparing forms and judging which is more primitive or advanced. His main criterion was what he previously called the "differentiation of functions and organs," but which he now identified with the concept of "division of labor," developed by Henri Milne-Edwards (who happened to be one of the judges at the French Academy).[48] Subsidiary measures of progress were independence, distinctness, and diversity of organs; reduction in the number of serially repeated segments or organs (with exceptions—loss of limbs or teeth or toes sometimes represented degeneration rather than progress); concentration and centralization of diffuse or serially repeated organs; and internalization and protection of surface features, such as gills or eyes.[49] These criteria were supposed to be derived from empirical observation of directional trends in the fossil record, but Bronn also seemed to think that the more advanced organisms were functionally or adaptively better off than the lower. Protecting eyes and gills, centralizing functions in specialized organs, and so on, were not only morphologically progressive, but also good ideas for safety and efficiency. Such considerations made it easy for Bronn later to accept Darwinian fitness as a concept related to his own idea of progress.

Other forms of progress discussed in *Geschichte der Natur* reappeared here as well, such as terripetal development (i.e., increasing numbers of freshwater and terrestrial species) and increasing numbers and diversity of species and other taxa, but Bronn now presented these trends more definitely as derivative. Bronn's fundamental law of adaptation, together with geological evolution, already put forth in *Geschichte der Natur*, now provided the ultimate explanation of terripetal development and other directional changes previously described as laws in their own right.

But Bronn's law of adaptation had done a better job of explaining the eliminative side of organic progress than the creative side. It had accounted for extinction of obsolete species and constrained the possibilities for the appearance of new ones, but it had not explained why or how the new species were produced. The version of 1858 now attempted to codify and simplify the laws of species production as well:

The succession of organisms from the very beginning of creation to the appearance of our present plant- and animal world was guided by two fundamental laws [*Grund-Gesetze*].

I. By an extensively, intensively, and constantly self-advancing, independent force of production.

II. By the nature of and the changes in the external conditions of existence, under which the organisms that are to be produced should live.[50]

Law I stipulated that the productive process would tend to improve itself with time and lead toward morphological progress, by Bronn's various criteria, such as physiological division of labor. Bronn did not claim that this law was logically or naturally necessary like the principle of adaptation, but he elevated it nonetheless to the level of *Grundgesetz*. It was now an equal partner with the principle of adaptation, which was implicit in Law II. Together these two laws saw to it that morphological progress and adaptation to a progressively changing environment occurred at the same time. All the other laws of progress emerged from the interplay of these two. For example, given the progressive availability of different environments over geological time, terripetal development resulted from the production of new and morphologically improved species under Law I, as modified by the need under Law II for adaptation to those diverse environments. Together the laws of progress and adaptation to a changing Earth yielded organic diversity.

As before, Bronn rejected older systems of embryological analogies or parallels between fossil sequences and embryonic stages. Bronn argued again that the direction of embryonic development was toward an increasingly perfect acquisition of its adaptations to adult life in a particular environment. Although this generally required the embryo to grow, differentiate, and increase in structural complexity, it would not necessarily perfect its form in any absolute sense or in parallel to any fossil sequence or scale of nature. Even if the embryo did happen sometimes to progress in similar ways to the fossil sequence, it would not do so consistently enough to imply any law of progress or recapitulation. The process of becoming adapted to adult life too often required nonprogressive, even regressive changes in the developing embryo. The number of stomachs in embryonic cows, for example, increased from one to four, which went against one of Bronn's criteria of progress: the reduction in the number of serially repeated organs. The number of the cow's toes decreased to two, which was below the mammalian norm and therefore also regressive. Still, both of these developments were in accordance with Bronn's law of adaptation, which was the higher, necessary one. Adaptation trumped progress, and Bronn tried to export this insight from paleontology to embryology.[51]

Bronn on the Laws and Forces of Morphology

Bronn did not dwell on embryology or comparative morphology in the prize essay, however. He made his historical argument quite independently of these traditional mainstays of zoology. Bronn set most of his material on morphology aside for separate publication, and it also came out in 1858, as *Morphologische Studien über die Gestaltungs-Gesetze der Naturkörper überhaupt und der organischen insbesondere* (Morphological studies of the formative laws of natural bodies in general and organic bodies in particular [hereinafter, "*Gestaltungs-Gesetze*"]).[52] The titles of his two 1858 works suggest that he was avoiding the single term *Entwicklung* for both historical and individual development, and was now using *Gestaltung* for present-day individual form.

In his *Gestaltungs-Gesetze*, Bronn revisited the unified, hierarchical system of laws and forces that he presented in 1841, and he aligned himself again with an ideal of scientific explanation based, like the Göttingen program, on laws and forces that unified biology with chemistry, mineralogy, geology, and physics. But Bronn had increased the number of laws and forces at work in determining both individual forms and the diversity of forms in the natural system, and he made form into a less-than-perfectly predictable product of their complex interactions.

Bronn enumerated four formative factors: Plan, *Entwickelung*, Adaptation, and *Manchfaltigkeit*. Plan was the basic arrangement of organs and parts that defined any of the major taxonomic groups. It was an extension of the Cuvierian concept of type, here analyzed into geometric and architectural components—the basic conoid or hemisphenoid shapes, the number of distinct organ systems, the number of repeated organs or segments, and the relative placement of important parts. The existence of many incommensurable body plans (not only at the level of the Cuvierian classes, but also in subgroups) provided for diversity, but without creating a hierarchy of types: no body plan was more perfect than any another.[53]

The laws of *Entwickelung* had to be superadded to the geometrical differences in order to account for progressive trends and the observed (intraclass) hierarchies of more or less perfect forms: "The second source, in contrast, develops the form-material created by the first, striding forward everywhere on definite pathways to unlike, permanent levels of organization, but without ever alienating an organism from its basic body plan."[54]

Thus the laws of *Entwickelung* created diversity within the major taxonomic groups, as some species rose higher than others. And since the laws and measures of progress were the same for all body plans, cross-group comparisons became possible, and some convergence in form could be explained. *Entwickelung* unified where Plan divided. Bronn also acknowledged an analogy between the progression of forms in embryonic development and the one observed in taxonomic arrangements, but did not extend the analogy to historical progress.[55]

Plan and *Entwickelung* by themselves did not guarantee a functional whole organism, because they were based upon purely internal considerations and made no reference to the environment in which the organism had to live. Bronn's third determinant of form, Adaptation, unified morphology with the physical sciences and saw to it that whatever the first two produced was capable of surviving under particular external conditions. Adaptation was a powerful force. It could trump *Entwickelung*, if necessary, and cause retrograde developments—in the embryo as well as in the taxonomic system. It could also account for convergences in form between species from separate Cuvierian groups. However, the convergences always proved to be superficial and could not alter the basic body plan.[56]

Finally, there was the problem of diversity. Bronn did not go into any detail in this work, but he admitted that the first three determinants of form, no matter how complex and unpredictable their interactions might be, did not by themselves account satisfactorily for every nuance of form. He therefore ascribed to nature a striving for diversity (*das Streben der Natur nach Manchfaltigkeit*), which made it play myriad variations on the themes given by Plan, *Entwickelung*, and Adaptation.

In his two essays of 1858, Bronn also reconsidered the problem of the origin of new species. He distanced himself from his earlier statements about divine intervention, and aimed for a more consistently naturalistic system:

The sober naturalist, who knows no force of nature that would produce plant and animal forms the way the force of attraction produces spherical world-bodies and the force of affinity shapes crystallized mineral species, will be inclined to view that force as flowing immediately from divine creative activity. However, that same sober naturalist will also say to himself that nothing else in nature is brought about by such a thing, but rather everything is organized and formed by general forces that are bound with matter. Therefore, here too, analogy definitely admonishes us to posit, even if it is as yet unknown to us, a similar force that brought forth and which perhaps, as Ch. Lyell assumes, now too, continues, although only very seldom, to bring forth plant- and animal species.[57]

Junker suggests that this shift away from an intervening God might have had more to do with meeting the expectations of the French prize committee than with an actual change of heart.[58] However, Bronn's 1858 stance is no temporary aberration. He again rejected the theistic solution in his 1860 critique of Darwin. Times had changed in Germany, too, and the French did not have the corner on atheistic and materialistic theories. Already in the 1840s, the theistic elements in *Geschichte der Natur* were unusually prominent for a German academic work, and out of character for one that gives an otherwise thoroughly physicalistic account of cosmology and planetary evolution. These elements became even more difficult to defend in the 1850s, given the ascendancy of physicalistic physiology and the raging *Materialismusstreit* (materialism controversy). This particularly nasty controversy broke out in 1854 at the annual meeting of the *Gesellschaft Deutscher Naturforscher und Ärzte* (*GDNÄ*, the German Society of Naturalists and Physicians)[59] and was still going strong as Bronn prepared his manuscript for the French Academy. Perhaps Bronn was seeking a naturalistic middle ground in his system of laws and forces, avoiding dogmatic assertions about either the material basis of life or the role of God in creation.

Despite having dropped his case for divine intervention, Bronn still expressed reservations about any naturalistic creative force. The problem was that its actions could not be observed in today's world, and it also appeared to have been inactive for intermittent periods in the past. Such erratic behavior conflicted with Bronn's uniformitarian conception of natural forces and processes.

Bronn used two strategies in 1858 to overcome his lack of a mechanism of species production. One was to study the effects of the productive process instead of its causes, and to continue, as before in *Geschichte der Natur*, to seek the patterns and laws of historical change. The other strategy was to divide the problem into two more-manageable parts: elimination of the obsolescent species and creation of the new ones. The eliminative side was governed as before by the law and necessity of adaptation. The creative side was less certain: "But even though the above laws undoubtedly obtain in Creation, we are far from maintaining that they [*with the exception of the negative side of the law of adaptation to external conditions*], are of such an absolute kind and effect as attraction, elective affinity, and other laws of nature that do not allow any exceptions"[60] (emphasis added, but brackets original, indicating an insertion by Bronn that was not in the French manuscript of 1855). Even worse, Bronn's account of the creative side lacked certainty not only about the very occurrence of progress, but its nature and direction: "We also do not in fact yet know which [taxonomic] compendium the Creator Himself consulted when prescribing the systematic sequence of creatures."[61]

Even more strongly, in the conclusion to *Entwickelungs-Gesetze*, Bronn called attention to this difference in the relative status of his laws, even as he touted his own accomplishments: "But the present treatise is rich in new results. It moves up the law of adaptation of the successive populations of the earth to the prevailing external conditions at any time to the position of the highest fundamental law [*oberstes Grund-Gesetz*]. In the negative direction, it is absolute; it allows no contradictory phenomenon, in the positive direction, however, it allows other, subordinate or independent, laws some room for play."[62]

The law of progressive development had a lower status, because progress could be delayed by the need to wait for proper geological conditions to appear, or compromised by the need to adapt a progressive design to prevailing conditions. From Bronn's methodological perspective, it was not an ideal solution, because the creative side of the succession problem was not brought under the rule of necessary and naturalistic laws. Conscious of his system's limitations, Bronn was still open to new suggestions. That was why he was prepared to read Darwin's *Origin* with such great interest and see it not as a surprising and revolutionary new proposal, but as a potential solution to his own remaining difficulties.

3

Darwin's *Origin*

As a British gentleman-naturalist, writing for other gentlemen like himself, Charles Darwin did not share all of Professor Bronn's concerns about the status of biology and paleontology as independent *Wissenschaften*. It was easier for Darwin to draw his disciplinary boundaries where he needed them, to bracket out the origin-of-life problem and the physical history of the Earth, and to decenter paleontology in his analysis of the past. Although had tried his hand at it in his work on the barnacles, compiling, systematizing, and illuminating large volumes of descriptive data was not Darwin's goal in the *Origin*, nor did he use existing systematic compilations as his main source of evidence for evolutionary change. Darwin made free use of anecdotal evidence and individual examples of variation and change, including personal observations from the *Beagle* voyage and a comparatively unsystematic assortment of crop plants, sheep, hunting dogs, and fancy pigeons. These latter examples were calculated to be familiar to British gentlemen, who might have bred some of them as a hobby or for use on their country estates, and they were crucial in the first chapter of the *Origin* for introducing basic concepts of variation, heredity, and artificial selection.

Unlike Bronn, Darwin was not looking for patterns in taxonomic or paleontological data from which to derive general laws of change. On the contrary, laws played a limited role in his theory, for Darwin agreed with the arguments of the natural theologians against the all-sufficiency of blind and rigid laws of nature. These alone could not fully account for complex adaptations and purposeful arrangements in nature. Compared to Bronn and other German morphologists, he would shift the balance of biological causes away from deterministic laws and toward historical contingencies and selective agents.

Still, Darwin did not arrive at his point of view in isolation. Darwin's studies and travels had exposed him to most of the major approaches of his time to the sciences of life. Thanks to mentoring from Robert Grant at the University of Edinburgh, Darwin had learned to appreciate earlier materialistic explanations of species transformation and progress, such as Lamarck's or that of his own grandfather, Erasmus Darwin, and possibly some of Meckel's ideas on embryology and evolution.[1] Also at Edinburgh, he encountered the last throes of the Neptunist–Vulcanist controversy and the newer catastrophism of Cuvier, as interpreted by the geologist Robert Jameson.[2]

Moving on from his medical studies at Edinburgh to his preparation for the clergy at Cambridge, Darwin studied the works of William Paley, the natural theologian, who focused his attention on the problem of explaining adaptation, but also of finding moral and religious lessons in nature. Darwin also assimilated theistic interpretations of nature from the geologist Adam Sedgwick and the botanist John Stevens Henslow, while deepening his knowledge of field natural history, catastrophist geology, and the progressive nature of historical change.[3] After his Cambridge connections helped to place him aboard the *Beagle*, he added Lyellian geology and Humboldtian biogeography to his repertoire of ideas and approaches and expanded his knowledge of species and their forms, adaptations, and distributions.[4]

After returning from the voyage, Darwin made the acquaintance of Richard Owen, who brought him up to date with German transcendental notions of archetypes, Baerian embryology, and Owen's own ideas about species succession, which were comparable to Bronn's, even if Bronn had been dismissive of them.[5] Darwin's notebooks from the late 1830s recorded his efforts to derive laws of adaptation, progression, and unity of type that were not far removed from Bronn's,[6] but he did not continue along these lines. Until 1859, he continually added new perspectives, deriving from, among other things, Malthus, Whig politics, competitive capitalism and investing, practical breeders, pigeon fanciers, and his eight-year study of the morphology, systematics, and paleontology of the barnacles.[7]

All things considered, then, there were many important points of convergence between Darwin and Bronn, particularly on the problems and methods of biogeography, on gradualism and uniformitarianism in geology and progress in the fossil record, and on the importance of adaptation—especially adaptation to the biotic environment—in explaining historical change. Were it not for this common ground, Bronn would hardly have taken such a lively interest in the *Origin*. But my focus in this chapter and the next is on the points of divergence between author and translator that will help explain Bronn's initial reaction, in his book review of 1860, and of course his spin on the translation and his critical commentary later that year. The most important of these divergences I attribute, directly or indirectly, to Darwin's Cambridge experiences and connections, and especially to his reading of Paley and his rejection of laws of adaptation, progress, and change such as Bronn's.

Paley and Darwin

During Darwin's time at Cambridge in the late 1820s, several of Paley's books, including *Natural Theology* and *Evidences of Christianity*, were still widely read (the latter even required), and they were favorites of young Darwin. As he wrote years later in his autobiography:

I am convinced that I could have written out the whole of the "Evidences" with perfect correctness, but not of course in the clear language of Paley. The logic of this book and as I may add of his *Natural Theology* gave me as much delight as did Euclid. The careful study of these works, without attempting to learn any part by rote, was the only part of the Academical Course which, as I then felt and as I still believe, was of the least use to me in the education of my mind. I did not at that time trouble myself about Paley's premises; and taking these on trust, I was charmed and convinced by the long line of argumentation.[8]

This connection between Darwin and Paley has been noted before, but I think the extent of Darwin's debt to Paley has been underestimated and overshadowed by historians' attention to Malthus, Lyell, and the context of industrialization and Whig politics. Also, the nature of Darwin's debt to Paley has not been fully explored. Paley's text explored more than just God's direct interventions into nature.

"In crossing a heath," Paley began his famous analogy, "suppose I pitched my foot against a *stone.*" He would have found that completely unremarkable and would not have troubled himself about how the stone came to be there or how it altogether came to be. "But," he continued, "suppose I had found a *watch* upon the ground." That would have required a special explanation because of the watch's evident purpose and intricate internal structure, whose intricacy, moreover, was clearly dedicated to serving the purpose. Paley's inference was, of course, that the watch was a designed object and must have had an intelligent designer and maker.[9] No other explanation would do, and the same reasoning applied *a fortiori* to living organisms, which were ever so much more complex and wonderfully purposeful than any human artifact:

Every indication of contrivance, every manifestation of design, which existed in the watch, exists in the world of nature; with the difference, on the side of nature, of being greater and more, and that in a degree which exceeds all computation. I mean that the contrivances of nature surpass the contrivances of art, in the complexity, subtlety, and curiosity of the mechanism; and still more, if possible, do they go beyond them in number and variety; yet, in a multitude of cases, are not less evidently mechanical, not less evidently contrivances, not less evidently accommodated to their end, or suited to their office, than are the most perfect productions of human ingenuity.[10]

Paley would not let the reader reserve judgment on design. He bombarded the reader with examples of organic contrivance and repeatedly challenged any atheist either to accept his Designer or to come up with a better explanation of their origins. He ruled out any alternative involving chance. He ruled out several other naturalistic alternatives as well, such as special laws or principles of order, or the effects of the animal's own striving to improve itself, and he repeated that there was no alternative to his conclusion.

Several connections have been drawn between Darwin and Paley in the historical literature. Paley focused Darwin's attention on the problem of explaining the origin of complex adaptations. He took a mechanistic view of life and drew analogies between physiological functions and feats of human engineering, foreshadowing Darwin's reliance on analogies to artificial breeding. His concept of Design applied not only to the organism, but to the entire balance of nature, where he found that Malthusian overproduction was not only universally at work, but also had beneficial effects. Darwin also seemed to rise to Paley's repeated challenges to find a better explanation of complex functionality than a Designer. As if to emphasize the fact that natural selection was meant to replace the designing God, Darwin imbued it—at least rhetorically—with many of the attributes of Paley's Designer, such as personality, powers of discernment, wisdom, and benevolence.[11] But there was even more to Paley and to Darwin's emulation of him, especially if one resists being hypnotized by the watch in Paley's allegory and keeps an eye on the stone as well.

Paley on Chance and Law

From the very beginning of his argument, Paley asked us to recognize designed objects like watches by their contrasts with other sorts of objects, like stones, and this implied that design was something special, not to be found everyone. Many natural objects or processes were like the stone, and did not require or receive the Designer's direct personal attention. Hence, Paley's God was not a micromanager, but allowed a variety of causes to operate in nature, more or less on their own. In all,

Paley discussed three kinds of causes: chance, law, and design. Each had a circumscribed role to play in the functioning of the world. The following analysis of Paley's *Natural Theology* treats it as a general account of natural causality, rather than only of design, and finds that Paleyan conceptions of chance and law were also very useful for Darwin.

Whatever else chance might have been good for, Paley rejected it out of hand as an explanation of things like watches: "Nor . . . would any man in his senses think the existence of the watch, with its various machinery, accounted for, by being told that it was one out of possible combinations of material forms; that whatever he had found in the place where he found the watch, must have contained some internal configuration or other; and that his configuration might be the structure now exhibited, viz. of the works of a watch, as well as a different structure."[12] Chance failed even more miserably as an explanation of a complex organ like the eye: "Something or other must have occupied that place in the animal's head," he imagined an atheist saying, in an effort to dismiss the whole problem. But of course, it was not just any old thing in the eye socket, but something highly unlikely: "That it should have been an eye . . . is too absurd to be made more so by any augmentation."[13]

Paley also took aim specifically at a commonly used materialistic explanation, the idea that an infinite universe, given infinite time and an infinite number of atoms in random motion, could have produced every possible configuration of matter, including functioning, reproducing organisms. Myriad other configurations would eventually have fallen apart again, but organisms would have multiplied and therefore persisted. The idea had a long history, dating back to the ancient atomists. It enjoyed a revival among eighteenth-century French materialists, and Paley caricatured Buffon's theory of organic particles that come together in a *moule intérieur* (inner form or mold) as just one more version of it (much as creationists today do with Darwinian variation and selection). In Paley's rendering: "The eye, the animal to which it belongs, every other animal, every plant, indeed every organized body which we see, are only so many out of the possible varieties and combinations of being, which the lapse of infinite ages has brought into existence; that the present world is the relict of that variety; millions of other bodily forms and other species having perished, being by the defect of their constitution incapable of preservation or of continuance by generation."[14]

Paley's objections to such schemes began with the fact that one did not observe nature experimenting and throwing new organisms together all the time, as required. Moreover, Paley argued, past experimentation should have produced all sorts of strange creatures that did not actually exist: "Upon the supposition here stated, we should see unicorns and mermaids, sylphs and centaurs, the fancies of painters, and the fables of poets, realized by examples." Or, in a more serious vein: "If it be alleged that these may transgress the limits of possible life and propagation, we might, at least, have nations of human beings without nails on their fingers, with more or fewer fingers and toes, than ten: some with one eye, others with one ear, with one nostril, or without the sense of smelling at all. All these, and a thousand other imaginable varieties, might live and propagate."[15]

These chance-based schemes also failed to explain the evident orderliness of the taxonomic system. If species were formed by the random collisions and aggregations of atoms, then, "a countless variety of animals might have existed, which do not exist." Species would not have fallen into "regular classes" as they clearly did: "The division of organized substances into animals and vegetables, and the distribution and sub-distribution of each into genera and species, which distribution is not an arbitrary act of the mind, but founded in the order which prevails in external nature, appear to me to contradict the supposition of the present world being the remains of an indefinite variety of existences; of a variety which rejects all plan."[16]

Paley did not say exactly how he thought species were distributed nonrandomly among genera, but Darwin looked into this matter and answered Paley's challenge in his chapter entitled "Variation in Nature." There, as will be discussed in detail below, he turned the tables on Paley by uncovering a surprising pattern that was not exactly random, but also not neatly organized the way one might have expected the divine Engineer to have ordained it.

Despite its inadequacy as an explanation of orderly and purposeful things, there was much that Paley did allow chance, defined now as the absence of intention, to accomplish in nature. Indeed, he found that there were some things that actually were *better* left to chance than to God's micromanagement or to perfectly regular organization:

Where order is wanted, there we find it; where order is not wanted, i.e., where, if it prevailed, it would be useless, there we do not find it. . . . In the forms of rocks and mountains, in the lines which bound the coasts of continents and islands, in the shape of bays and promontories, no order whatever is perceived, because it would be superfluous. No useful purpose would have arisen from moulding rocks and mountains into regular solids, bounding the channel of the ocean by geometrical curves; or from the map of the world, resembling a table of diagrams in Euclid's Elements, or Simpson's Conic Sections.[17]

Paley seems to have assumed that disorder was the original state of things from the Biblical Beginning, when all was *tohu vavohu*. The stone that he accidentally kicked in his opening allegory, he said, might always have been just as it was and where it was, and that was all he cared to know about it. Similarly, rocks, coastlines, and so on also retained their original disorder. Only departures from this original disorder required any explanation other than the operation of chance.

Chance was operative not only in the physical world and in things that have lain around since the Creation, but also in living things and even in human affairs. Simple, unimportant or pathological features of organisms may have been shaped or placed by chance: "What does chance ever do for us? In the human body, for instance, chance, i.e. the operation of causes without design, may produce a wen, a wart, a mole, a pimple, but never an eye. Amongst inanimate substances, a clod, a pebble, a liquid drop might be; but never was a watch, a telescope, an organized body of any kind, answering a valuable purpose by a complicated mechanism, the effect of chance. In no assignable instance hath such a thing existed without intention somewhere."[18]

But the Designer might also have let chance determine more consequential outcomes. In human affairs, Paley pointed out by way of analogy, it was often fair and expedient to draw lots, for example when assigning tasks to members of an organization or when distributing favors. In the design of individual humans, God, too, needed a fair method of assigning lots in life—wealth, status, talents, diseases, constitutions, and longevity. Since He loved and judged everyone equally, He would not have wanted to play favorites and intentionally give one soul a better material existence than another. But on the other hand, a perfectly equitable distribution would not have been a good design, because the ship of state would not have been able to sail very well had it not been manned by a crew of diverse individual constitutions, talents, and stations. So what better, fairer solution was there than for the Designer to step back and let material and bodily advantages be distributed by chance?[19]

Darwin's concept of random variation was more than just foreshadowed in this Paleyan distribution of lots in life. Darwinian variation was undesigned and represented no divine favor, disfavor, or plan. No purpose could be ascribed to it, and there was little point in seeking its causes. In addition, Paley even implied that randomly assigned characteristics affected an individual's ability to function in the economy of the state, just as Darwinian variations affected an individual's ability to function in the economy of nature.

In one small but suggestive aside, Paley also described how undesigned raw materials could be put to use for some purpose, which was what Darwin needed to happen under natural selection. A craftsman could do this quite easily, Paley said: "A cabinet-maker rubs his mahogany with fish-skin; yet it would be too much to assert that the skin of the dog-fish was made rough and granulated on purpose for the polishing of wood, and the use of cabinet-makers." Paley minimized the importance of such instances in nature, however, on the grounds that it was too much to expect from organisms: "Is it possible to believe that the eye was formed without any regard to vision; that it was the animal itself which found out, that, though formed with no such intention, it would serve to see with?"[20] Darwin had to find another agency than either God or the organism itself that would figure out what to do with nature's undesigned resources.

Paley's third major category of natural cause comprised the laws or other blind and inflexible organizing and formative principles of nature, and these, too, had their strictly circumscribed realm of operation. Except when used by the Designer as tools, the laws of nature could not form anything complex and purposeful. Without divine guidance, laws and other organizing principles would have created useless and repetitious regularities, never nature's complex web of interrelated organs, organisms and environments. Approaches such as Kielmeyer's or (later) Bronn's, requiring laws of production, adaptation, or progress, were ruled out as a matter of principle.

The watch in Paley's analogy, therefore, was no better accounted for by law than by chance. No man in his senses would believe "that there existed in things a principle of order, which had disposed the parts of the watch into their present form and situation. He never knew a watch made by the principle of order; nor can he even form to himself an idea of what is meant by a principle of order, distinct from the intelligence of the watch-maker."[21]

That same sensible man would also have been surprised to hear "that the watch in his hand was nothing more than the result of the laws of *metallic* nature. It is a perversion of language to assign any law, as the efficient, operative cause of any thing. A law presupposes an agent; for it is only the mode, according to which an agent proceeds; it implies a power; for it is the order, according to which that power acts. Without this agent, without this power, which are both distinct from itself, the *law* does nothing; is nothing" (emphasis in original).[22]

Moving from the artificial to the natural, Paley found it just as absurd to attribute organic design to the operation of unguided laws and their unintended consequences: "The expression, 'the law of metallic nature,' may sound strange and harsh to a philosophic ear; but it seems quite as justifiable as some others which are more familiar to him, such as 'the law of vegetable nature,' 'the law of animal nature,' or indeed as 'the law of nature' in general, when assigned to the cause of phenomena, in exclusion of agency and power; or when it is substituted into the place of one of these."[23]

Paley acknowledged that Buffon's *moule intérieur* might have been counted as an organizing principle that arranged the organic particles, but that did not help Buffon's case, in Paley's opinion. No one could tell what such a thing really was or how it worked. It was just an empty word. Similarly, Paley also discussed and dismissed the idea of self-organizing matter, probably with Erasmus Darwin in mind, but the same could have applied to Lamarck as well: "The principle, and the short account of the theory, is this. Pieces of soft, ductile matter, being endowed with propensities or appetencies for particular actions, would, by continual endeavours, carried on through a long series of generations, work themselves gradually into suitable forms." By such reasoning, Paley continued, one might have expected that "a piece of animated matter . . . that was endued with a propensity to *fly*, though ever so shapeless, though no other we will suppose than a round ball to begin with, would, in a course of ages, if not in a million years, perhaps in a hundred millions of years (for our theorists, having eternity to dispose of, are never sparing in time), acquire *wings*"[24] (emphasis in original). Paley argued that such propensities for self-improvement or self-organization, if they were in operation at all, must have been imposed upon living matter by the Designer and were only His tools, not independent causes. Such purposeful qualities would not have arisen spontaneously.

Still, there was a place in Paley's system for law-governed actions, and one of them was of immense importance for Darwin: Malthusian superfecundity or overproduction. This was an example of a predictable, law-abiding "power" with which the Designer endowed all the animals, and which ran its course without His direct intervention. Even though it had the undesirable side effects of continual overpopulation, misery, and loss of life, Paley had no trouble finding the potential for good in it, just as Darwin later made overproduction into a force for evolutionary progress.

Paley believed that God's creatures were generally happy, despite the pressures of competition and depredation. Both predator and prey took pleasure in the bodily exertion of the chase, and when the time came for the prey to succumb, death was swift and painless. Superfecundity also assured that the total amount of happiness in the world remained as high as possible, by producing as many happy creatures as could possibly fit. Wherever there might be a "vacancy fitted to receive" greater numbers of some species, superfecundity made it possible for that species to fill it up happily: "It pours in its numbers, and replenishes the waste."[25]

Once nature was full again, countervailing forces of destruction, such as predators, diseases, and hunger, prevented overgrowth. Such destructive forces were not evil, but rather part of the grand design, which was certainly good:

But then this *superfecundity*, though of great occasional use and importance, exceeds the ordinary capacity of nature to receive or support its progeny. All superabundance supposes destruction, or must destroy itself. Perhaps there is no species of terrestrial animals whatever, which would not overrun the earth, if it were permitted to multiply in perfect safety; or of fish, which would not fill the ocean: at least, if any single species were left to their natural increase without disturbance or restraint, the food of other species would be exhausted by their maintenance. It is necessary, therefore, that the effects of such prolific faculties be curtailed. In conjunction with other checks and limits, all subservient to the same purpose, are the *thinnings* which take place among animals, by their action upon one another.[26] (Emphases in original)

Thus, long before his celebrated reading of Malthus in 1838, Darwin had been primed by Paley to think that it was beneficial for species to overreproduce to the point of falling victim to disease, depredation, and want, because it let them capitalize quickly on vacancies in nature. He had also read that, at least in the case of humans, individuals were not all exactly alike, and that their constitutions and talents were assigned at random. Darwin's additional insights of 1838 were the generalization from this to the chance occurrence of beneficial variations of all sorts, and the link between variation and success during periods of overpopulation and destruction.

In contrast, Bronn had also read his Malthus and applied it to plants and animals, but not from a Paleyan perspective. He described geometric increase and gave some examples in his *Geschichte der Natur*, and he inferred that as a result of superfecundity the observed balance of nature could only have been maintained at the cost of many individual lives. Hence, according to Bronn, survival in nature involved continual struggle. That was the necessary outcome, and Bronn offered no judgment about whether it was beneficial to the affected organisms. If anything, he emphasized the destructive role of struggle in causing extinction when a species was no longer adapted to prevailing conditions. He did not write about opportunities for expansion that might become available to fecund species. He associated superfecundity and struggle with extinction, and he looked for laws of another sort to account for the creation of species and their adaptations.

Variation Under Domestication and in Nature

Let us now consider Darwin's approach in greater detail, with an eye toward its Paleyan features and its appeal to British gentlemen-naturalists. In his "Introduction" to *The Origin of Species,* Darwin identified and positioned himself first and foremost as a British field naturalist. He opened a with a reference to his voyaging days and his observations on the biogeography of living and fossil organisms, which seemed to demand an evolutionary explanation: "When on Board H.M.S. 'Beagle,' as naturalist, I was much struck with certain facts in the distribution of the inhabitants of South America, and in the geological relations of the present to the past inhabitants of that continent. These facts seemed to me to throw some light on the origin of species—that mystery of mysteries, as it has been called by one of our greatest philosophers."[27] Alfred Russel Wallace and Joseph Hooker were cited prominently, further underscoring Darwin's concerns with biogeography and scientific travel. Issues in morphology were given much less weight, and there was no grand discussion of scientific method or the place of biology among the sciences.

Before introducing any specifics about his theory, Darwin made an important concession to the natural-theological point of view. Even though many facts of geographic distribution, fossil succession, morphology, and systematics might have seemed to point toward a theory of species transformation, he argued, the real problem that needed to be addressed was still the old one singled out by Paley, namely, how to explain complex, purposeful structures and adaptations: "Nevertheless, such a conclusion [i.e., that species may become transformed], even if well founded, would be unsatisfactory, until it could be shown how the innumerable species inhabiting this world have been modified, so as to acquire that perfection of structure and coadaptation which most justly excites our admiration."[28]

Darwin thus aligned himself with his old Cambridge colleagues, and presumably kept them turning the pages, by emphasizing his rejection of evolutionary theory as they knew it, using familiar Paleyan arguments against it. First he dispatched theories that relied, like Lamarck's or

Erasmus Darwin's, on environmental effects, use and disuse of organs, or the will or the perceived needs of the organism:

It is preposterous to attribute to mere external conditions, the structure, for instance, of the woodpecker, with its feet, tail, beak, and tongue, so admirably adapted to catch insects under the bark of trees. In the case of the misseltoe, which draws its nourishment from certain trees, which has seeds that must be transported by certain birds, and which has flowers with separate sexes absolutely requiring the agency of certain insects to bring pollen from one flower to the other, it is equally preposterous to account for the structure of this parasite, with its relations to several distinct organic beings, by the effects of external conditions, or of habit, or of the volition of the plant itself.[29]

For the same reason, Darwin also rejected law-based theories of evolution, among which he singled out the one proposed by Robert Chambers in *Vestiges of the Natural History of Creation*.[30] Drawing in part on German sources, Chambers had argued that the laws of generation need not always have made like produce like, but might have been programmed to produce, over the long run, a progressively varying sequence of forms: "The author of the 'Vestiges of Creation' would, I presume, say that, after a certain unknown number of generations, some bird had given birth to a woodpecker, and some plant to the misseltoe, and that these had been produced perfect as we now see them; but this assumption seems to me to be no explanation, for it leaves the case of the coadaptations of organic beings to each other and to their physical conditions of life, untouched and unexplained."[31] Such blind operations of fixed laws might indeed have produced a changing sequence of forms, but as Paley said, it would have been an arbitrary one. The forms could hardly have been expected to coincide so perfectly in time and place with environmental changes and with changes in other organisms as to result in complex interdependencies.

With Lamarck and law ruled out, and chance not even deemed worthy of consideration, what other candidates were there for the cause of adaptation? It had to be a new kind of cause that was not on Paley's list but would potentially answer Paley's challenge and explain complex adaptations. Selection was just such a new kind of cause, one that could do all that Paley's Designer did to shape and adapt organisms. To introduce selection, Darwin began in chapter 1 with the human craft of artificial breeding, before going on to argue, Paley-like, that there must have been something similar, but more powerful, at work in nature.

There were some very close correspondences between artificial and natural selection. Not only was artificial selection a cause of organic change generally, but it produced some of the same patterns of change that Darwin wished to explain in nature. It accounted for "adaptations" of sorts, if one was willing to count conformity to the breeder's specifications as an analogue of what was needed for survival in nature. It showed that change by selection was gradual and cumulative over many generations. It accounted for the phenomenon of divergence of lineages, for example when different breeders or fanciers, who typically "will not admire a medium standard,"[32] selected for different extremes in a range of variations. It also illustrated how intermediate forms easily could be lost, or not even produced, in such cases obviating the need for Darwin to produce numerous transitional fossils or to apologize for gaps in the animal series. Artificial selection was even shown to operate "unconsciously," when the farmer simply bred his best plants or animals, without any goal or design in mind for modifying them or producing a new variety.

There was only one small difference that Darwin admitted indirectly, perhaps even unintentionally, by means of the words he used to describe the direction of change in nature and under domestication. Varieties or species in nature were said to be "improved" (the word was used 45 times in this sense) or "perfected" (16 times) by natural selection. Domestic varieties were invariably said to be "improved" (19 times) by artificial selection, and never "perfected."[33] As discussed further in the next chapter, these appeared to Bronn to be important distinctions. He magnified the difference in the translation, and related it to the differences between scales of morphological progress.

Here in Darwin's *Origin*, the frequent usage of "improved" for both artificial and natural outcomes suggested that selection *usually* was making the same kinds of modifications in both cases. But why was Darwin not consistent in calling the natural outcome "improvement"? Were the two terms meant to distinguish among improvements in suitability for human purposes, competitive ability or fitness, and advancement up a scale of morphological perfection? I think rather that he was trying to conflate these different directions of change. When Darwin wrote "perfected," he was addressing the morphologists in their own language but subtly changing the meaning of their term at the same time. By using it interchangeably with "improved" in the context of natural selection, he linked morphological advancement with competitive improvement as well as with artificial improvement. Contrariwise, Darwin might perhaps have been read as implying that the correspondence between artificial and natural selection did not hold in every detail and that there were times when the word "improvement" did not quite capture all that natural selection was capable of accomplishing. There was still something in nature that deserved to be called "perfection" and which artificial selection did not produce. In the German *Origin*, Bronn opted for this latter interpretation.

Darwin's analogy between artificial and natural selection has been seen as device for establishing the efficacy of selection and its legitimacy as a *vera causa*, or "true cause" in the sense of John Herschel,[34] the philosopher to whom Darwin alluded as "one of our greatest." The term comes from Newton's first rule of reasoning, that "We are to admit no more causes of natural things than such as are both true and sufficient to explain their appearances,"[35] and Darwin was evidently trying to satisfy two competing interpretations of this rule, by Herschel and William Whewell. Herschel demanded that any proposed new cause be analogous to known natural causes, the way that Newton's gravitational force had been analogous to other sorts of mechanical forces. This had the effect of reining in the imagination and avoiding the kind of hypothesizing to which Newton famously had been opposed. A related methodological consideration was that natural selection had to satisfy the demands of uniformitarianism for the use of processes observable in the present, as explanations of the past. From that point of view, artificial selection provided a present-day equivalent to past selection in nature. Whewell, in contrast to Herschel and the uniformitarians, placed no such advance restrictions on natural causes, but proposed to recognize the true ones by their "consilience," or ability to provide unifying explanations of phenomena previously thought to be unrelated.

The analogy showed that selection, whether artificial or natural, was admissible by Herschel's standard as a true natural cause and not a wild hypothesis. As Darwin wrote, "The great power of this principle of selection is not hypothetical. It is certain that several of our eminent breeders have, even within a single lifetime, modified to a large extent some breeds of cattle and sheep."[36] Once he had established selection as *a* true cause, he could go on to make the case that it was also *the* true cause of the effects in question, and even unify many effects with it, as required by Whewell.

German biologists, in contrast, had exhibited little interest in the *vera causa* problem and seldom wrote about it, even though, they, too, took Newton as their model for scientific method. They proceeded directly to Newton's second rule of reasoning, which said to assign the same cause to the same effects. This was central, for example, to Kielmeyer's methodology, when he grouped physiological effects, tied each group to a single cause, and tentatively called it a "force." There was no hand-wringing over the admissibility of such forces as causal explanations, whether they were true causes or hypothetical ones, or what they even were. The important thing was to uncover and if possible quantify the laws that governed their actions, which was what Bronn expected from Darwin, too.

With the argument from artificial selection, Darwin not only addressed Herschel's expectations, but also took a page out of Paley's *Natural Theology*, which likewise began with an analogy to human artifice. These uses of artificial selection seemed calculated to win over a British audience, especially since Darwin illustrated the chapter with familiar British breeds of plants and animals. To appreciate the chapter fully, one would almost have to have grown up, like Darwin, reading Paley and being "charmed and convinced by the long line of argumentation," followed developments in British philosophy, and been familiar with British sheep, hunting dogs, and fancy pigeons. For a German academic reader like Bronn, the analogy to artificial selection was of much less interest and cogency.

Chance, Design, and a Manufactory of Species

Darwin's treatment of variation and selection, whether under domestication or under nature, followed Paley in more than just the drawing of an analogy between the artificial and the natural. It also applied Paley's conception of order and purpose as qualities that had to be imposed by some agency upon disorderly features of nature. Darwinian variation, like the warts and wens and stones in Paley's work, were formed by chance, that is, without order or purpose, before being put to use by artificial or natural selection. This is a point that Darwin insisted upon repeatedly to his theistically inclined supporters, Charles Lyell and Asa Gray, who wanted to see divine purpose in the occurrence of favorable variations. As Darwin wrote to Gray: "I have lately been corresponding with Lyell, who, I think, adopts your idea of the stream of variation having been led or designed. I have asked him (& he says he will hereafter reflect & answer me) whether he believes that the shape of my nose was designed. If he does I have nothing more to say. If not, seeing what Fanciers have done by selecting individual differences in the nasal bones of Pigeons, I must think that it is illogical to suppose that the variations, which Nat. Selection, [*sic*] preserves for the good of any being, have been designed."[37]

In his book on *Variation Under Domestication*, Darwin switched from noses to the more conventionally Paleyan example of stones to make the same point, which was that the world was full of originally undesigned things that eventually could be put together to serve a purpose: "Let an architect be compelled to build an edifice with uncut stones, fallen from a precipice. The shape of each fragment may be called accidental; yet the shape of each has been determined by the force of gravity, the nature of the rock, and the slope of the precipice,—events and circumstances, all of which depend on natural laws; but there is no relation between these laws and the purpose for which each fragment is used by the builder."[38]

Moving again from artifice to organism, Darwin equated variation with the stones and selection with the architect: "In the same manner the variations of each creature are determined by fixed and immutable laws; but these bear no relation to the living structure which is slowly built up through the power of selection, whether this be natural or artificial selection. If our architect succeeded in rearing a noble edifice, using the rough wedge-shaped fragments for the arches, the longer stones for the lintels, and so forth, we should admire his skill even in a higher degree than if he had used stones shaped for the purpose."[39]

This was also the reason why Darwin did not need to trouble himself too much about the precise causes of variation: whatever those causes might have been, they were not the causes of the resulting complex adaptations. Selection was the cause, which, like Paley's Designer, had to impose purpose on the raw materials that undesigned nature provided: "The key is man's power of accumulative selection: nature gives successive variations; man adds them up in certain directions useful to him. In this sense, he may be said to make for himself useful breeds."[40]

To meet Paley's challenge to explain complex, purposeful structures and functions without recourse to design, Darwin needed to show that Paley had not exhausted the possibilities with his dismissive treatment of chance and law as explanations. Selection filled this need, because it amounted to a new kind of natural cause that was neither a matter entirely of chance, nor of lawlike necessity, nor of design. Darwin applied this new kind of cause for the first time in chapter 2, together with his first Paley-like challenge to an imaginary opponent to come up with a better explanation of a phenomenon.

Recall that Paley had argued that the orderliness of the taxonomic system, in particular the distribution of species among the genera, defied explanation by any atheistic scheme for generating species at random.

But how orderly was this distribution really? On the basis of his own theory, Darwin predicted a partially ordered arrangement, featuring a hierarchy of groups within groups, but with highly variable and unpredictable numbers of subgroups per group. The result looked like a lopsided tree, in which the groups or branches that happened to have been most successful (in terms of numbers of individuals and geographic dispersal) had generated the most subgroups or twigs. Success led to success: "From looking at species as only strongly-marked and well-defined varieties, I was led to anticipate that the species of the larger genera in each country would oftener present varieties, than the species of the smaller genera; for wherever many closely related species (*i.e.*, species of the same genus) have been formed, many varieties or incipient species ought, as a general rule, to be now forming. Where many large trees grow, we expect to find saplings."[41]

Chambers, as a proponent of the rule of law, would never have expected such an outcome. He had endorsed the quinarian system—a perfectly regular hierarchy in which each group always had five subgroups. Exactly how a Paleyan Designer would have distributed species and varieties was not clear. Paley had not specified a pattern, and Darwin did not attribute any particular one to him. Paley could easily have had a linear scale of nature in mind. Some perfectly regular hierarchical arrangement like the quinarian might also have suited his argument against the efficacy of chance. But there was no reason to expect the Designer to have created lopsided taxonomic trees: "If we look at each species as a special act of creation, there is no apparent reason why more varieties should occur in a group having many species than in one having few."[42]

Darwin undertook a quantitative study of the distribution of species and varieties among the genera, based on standard compendia of British plants and beetles, and he presented his results as a test of his prediction. Sure enough, species in the larger genera (those with more than the median number of species per genus) were more likely to present varieties than species in the smaller genera, and when they did, they presented on the average a larger number of them. Darwin tried the same test again, omitting the very smallest genera (those with fewer than four species), and again got the same result. Like Paley's hapless athcist, Darwin's creationist had no way to predict or explain this result. The argument might seem like one of Darwin's more obscure ones, but as we shall see in chapter 5, it was the one that first impressed Ernst Haeckel, because he found that it helped him account for patterns of variation that he was finding in his classification of the Radiolaria.

At this point in his book, Darwin had not yet detailed the operations of struggle and natural selection and did not use this result as evidence for them. For the time being, it was only a question of the *pattern* of variety and species production, not the mechanics of it. Darwin did not yet write of natural selection, but of a "manufactory" of varieties and species, which evidently operated on its own, without constant guidance, and produced its most salable wares in more and more versions: "These facts are of plain signification on the view that species are only strongly marked and permanent varieties; for wherever many species of the same genus have been formed, or where, if we may use the expression, the *manufactory* of species has been active, we ought generally to find the *manufactory* still in action, more especially as we believe the process of *manufacturing* new species to be a slow one"[43] (emphasis added).

It is significant that the cause of species production was described as a manufactory, rather than a law or a mechanism. It was not like the laws of Newtonian mechanics, because it did not predict individual outcomes. Darwin could not say when or precisely why a manufactory might become active in any region. It was also not random, like the collisions of atoms in the old materialistic systems, because it produced a consistent distribution of species in taxonomic groups within taxonomic groups. It was a new category of natural cause, one that Paley had not considered and rejected, and one that could explain things that neither chance nor law nor design could account for.

Struggle and Selection, and Explanation

In chapter 3 of the *Origin*, Darwin began to detail how the manufactory took in individual variations and made them into varieties, species, genera, and higher categories. The process began with a struggle for existence, brought on inevitably by Malthusian superfecundity, but Darwin approached the matter of struggle in a roundabout way. Sounding almost like a natural theologian, Darwin first emphasized nature's bounty and the happiness of her creatures: "We behold the face of nature bright with gladness, we often see superabundance of food." But then he reminded his readers that there were some unpleasant realities beneath the superficial happiness: "We do not see, or we forget, that the birds which are idly singing round us mostly live on insects or seeds, and are thus constantly destroying life; or we forget how largely these songsters, or their nestlings, are destroyed by birds and beasts of prey; we do not always bear in mind, that though food may now be superabundant, it is not so at all seasons of each recurring year."[44]

Still, this was not as radical a departure from Paley's optimistic view as it is sometimes made out to be, for Paley did not omit to mention the underlying destruction of life. He had described the harsh checks on overproduction as the price that had to be paid for its benefits. Paley had described how superfecundity allowed happy creatures to capitalize on opportunities in nature and to fill any vacancies, and Darwin now described a similar process, albeit in much harsher language and with a less literal interpretation of what constituted a "vacancy" in nature: "Lighten any check, mitigate the destruction ever so little, and the number of the species will almost instantaneously increase to any amount. The face of nature may be compared to a yielding surface, with ten thousand wedges packed close together and driven inwards by incessant blows."[45] With Darwin, vacancies in the economy of nature did not just happen to open up. Species had to make their own opportunities at the expense of other species, "wedging" them out of the way.

Even for Darwin, everything was for the best—usually. There was neither a benevolent Designer nor a law of nature that guaranteed it, but generally, the competing wedges were so equally matched that on the whole, the balance of nature was maintained, and the bright face of nature was not so deceptive after all: "Battle within battle must ever be recurring with varying success; and yet in the long-run the forces are so nicely balanced, that the face of nature remains uniform for long periods of time, though assuredly the merest trifle would often give the victory to one organic being over another."[46] The struggling individuals could still enjoy life: "When we reflect on this struggle, we may console ourselves with the full belief, that the war of nature is not incessant, that no fear is felt, that death is generally prompt, and the vigorous, the healthy, and the happy survive and multiply."[47]

In chapter 4, on natural selection, Darwin's language suggested most strongly that selection was intended to play the role of Paley's designer.[48] He emphasized that it was just like a human craftsman, only much better: "As man can produce and certainly has produced a great result by his methodical and unconscious means, what may not nature effect?"[49]

Natural selection also had all the attributes of Paley's Designer. Natural selection was all-seeing: "Man can act only on external and visible characters: nature cares nothing for appearances, except in so far as they may be useful to any being. She can act on every internal organ, on every shade of constitutional difference, on the whole machinery of life."[50]

Natural selection was benevolent, and like a shepardess with her flock, she attended to the needs of each individual, so that it could develop its adaptive characteristics to the fullest: "Man selects only for his own good; Nature only for that of the being which she tends. Every selected character is fully exercised by her; and the being is kept under well-suited conditions of life."[51]

Natural selection also had such endless skill and patience that it could produce the very best results: "How fleeting are the wishes and efforts of man! How short his time! And consequently how poor will his products be, compared with those accumulated by nature during whole geological periods. Can we wonder, then, that nature's productions should be far 'truer' in character than man's productions; that they should be infinitely better adapted to the most complex conditions of life, and should plainly bear the stamp of far higher workmanship?"[52]

Furthermore, natural selection intervened personally in evolution and exercised judgment, much more like the Deity than a fixed law of change: "It may metaphorically be said that natural selection is daily and hourly scrutinizing, throughout the world, every variation, even the slightest; rejecting that which is bad, preserving and adding up all that is good; silently and insensibly working, whenever and wherever opportunity offers, at the improvement of each organic being in relation to its organic and inorganic conditions of life."[53]

This might have been effective rhetoric to use on British audiences, who shared Darwin's admiration of Paley and might have expected any proposed cause of adaptation to play all of the Designer's roles. Unfortunately, the more natural selection sounded like a personal agent, imposing order upon random variation, the less it sounded to Bronn like an appropriate explanatory device in a *Wissenschaft* of life. Bronn was expecting a system of laws and forces that would reflect the inherent orderliness of nature, maintain the integrity of species and taxonomic categories, and produce the observed sequences of forms. Instead he was getting a chaos of variations and an unpredictable, anthropomorphic meddler sometimes imposing a kind of order.

Bronn tried hard to overlook the anthropomorphism and to view natural selection as a law or force of nature, as he went through the remaining chapters of the *Origin*. In those chapters, Darwin first enumerated the difficulties that he thought would be raised, and showed that he could explain most if not all of them away. He then triumphantly applied natural selection and descent with modification to a wide range of natural phenomena that had been troublesome for either natural theology or successional theories like Lyell's. Chief among the problems were those striking facts of South American biogeography and paleontology with which the book began, and which were only striking because they defied any conceivable logic of Design.

To take just one example of many, why was it that when one traveled southward in South America and the climate got colder, the common rhea was replaced by the lesser rhea? Why another rhea, rather than some other large, flightless bird, designed from scratch for the new environment? In contrast, when moving eastward from South America, across the oceans, one could stay in the same climate zone and encounter no more rheas, but ostriches in Africa and emus in Australia. Why the new designs for the same sorts of environments? Which, if any, was the perfect flightless bird for that climate zone? Similarly, why did Darwin find fossils of giant armadillos in the same regions where armadillos live today? If environmental change drove the giants to extinction, why would the Creator or creative force have replaced them with smaller versions and not something else entirely—again something perfect for the new environment?

Darwin argued that there must have been something that kept the armadillos together over time, the way the rheas were kept together over contiguous continental space: "We see in these facts some deep organic bond, prevailing throughout space and time, over the same areas of land and water, and independent of their physical conditions. The naturalist must feel little curiosity, who is not led to inquire what this bond is."[54] Darwin had a simple answer:

This bond, on my theory, is simply inheritance, that cause which alone, as far as we positively know, produces organisms quite alike, or, as we see in the case of varieties, nearly like each other. The dissimilarity of the inhabitants of different regions may be attributed to modification through natural selection, and in a quite subordinate degree to the direct influence of different physical conditions. . . . On this principle of inheritance with modification, we can understand how it is that sections of genera, whole genera, and even families are confined to the same areas, as is so commonly and notoriously the case.[55]

Darwin could explain much more than those striking facts of distribution and succession that inspired him to question his early Paleyan assumptions about design. He took on problems of instinct, social behavior, parasitism, hybrid sterility, classification, embryonic development, vestigial organs, and homologies, and he solved them all with alacrity, while also hinting that there was much more to be done with human history and psychology. These disparate fields of biological inquiry became unified within the Darwinian system of historical explanations, which therefore exhibited consilience, in the sense of the Whewellian *vera causa*.

The advantages of such unifying explanations were not lost on Bronn, even though he had no explicit concerns about *verae causae*. The problem he had was a feeling of being left out of a conversation that Darwin was carrying on with an imaginary Lyellian successionist or a natural theologian. For example, Darwin asked why an oceanic island like any of the Galapagos should have a fauna similar to that of the nearest continent, instead of a unique one for its unique environment, or one more like the fauna of similar oceanic islands in other parts of the world, like Cape Verde. The fact was awkward for the natural theologian, because it suggested that the design requirement for adaptation did not fully determine what species were placed where. Darwin declared, in the style of Paley, that his own view provided the only viable explanation: "I believe this grand fact can receive no sort of explanation on the ordinary view of independent creation; whereas on the view here maintained it is obvious that the Galapagos Islands would be likely to receive colonists, whether by occasional means of transport or by formerly continuous land, from America; and the Cape de Verde Islands from Africa; and that such colonists would be liable to modification."[56] Darwin repeated this Paleyan tactic of weighing competing explanations throughout the *Origin*, and I interpret it as both an influence of Paley's logic and a tactic for arguing with the Cambridge- and Oxford-educated gentlemen who were Darwin's closest associates.

From Bronn's point of view, however, this tactic was ineffective, or worse. Not having grown up with Paley, Bronn was unimpressed by the form of the argument, and did not feel compelled to equate a good explanation with a true one. He could also see that Darwin was playing the game of competing explanations without a full deck, for Bronn's own favored explanation was not among the competitors that Darwin weighed against natural selection. Bronn's brand of successionism was not committed to perfect design and perfect geographical placement. Organic forms, for Bronn, were produced by the joint influences of local conditions, the law of adaptation, structural considerations related to type, and various laws of diversification and progress. Adaptation had to be good enough for survival, to be sure, but not necessarily perfect. Bronn's system could accommodate any number of design compromises. And in any case, Bronn posited a dynamic environment, so no matter how perfectly adapted a species might have been when it was created or otherwise generated, progressive changes of the Earth's surface gradually reduced its adaptedness and diminished its numbers until it eventually went extinct.

This particular argument of Darwin's about the illogic of island biogeography also left Bronn unscathed. If island faunas consistently resembled those of the nearest mainland, that only went to show the importance of geographic location in influencing the creative force. So, although Bronn was impressed by Darwin's ability to supply good explanations of many disparate phenomena, he did not see Darwin explicitly challenging his own explanations of those same phenomena.

There were also many features of Bronn's work that Darwin did not even attempt to bring under his system of consilient historical explanations. Darwin never validated his theory against detailed tables of fossil occurrences or compilations of morphological descriptions like Bronn's. He avoided committing himself to particular evolutionary histories of particular groups. Darwin never dealt with the origin and the nature of life, its relationship to chemicals and minerals, or the causes and patterns of geological change. Worst of all, Darwin implied that naturalists would have to construct an individualized, hypothetical story of migration, environmental change, variation, struggle, selection, and modification for each and every species or lineage. It was easy to do this, indeed too easy, for Bronn's taste, for there was little danger of having the stories proven wrong. Under Darwin's theory, hypotheses multiplied, and no universal patterns and laws of change were discovered.

Also lacking in Darwin's book was any analysis of what was meant morphologically by "perfection" or "improvement," comparable to Bronn's efforts to define *Vervollkommnung*. Most of the time, the direction of evolutionary change was clearly supposed to be toward better competitive ability, but as Bronn complained, there was no way to know exactly what made the superior competitor better, and no way to compare and rank species without setting one loose in the other's territory. Sometimes Darwin indicated that competitive superiority had something to do with increasing efficiency and division of labor.[57] As we shall see in the next chapter, Bronn did not criticize this ambiguity as a deficiency in Darwin, but rather used it as an opportunity to insert his own analysis. He tried to link competitive ability, by way of division of labor, to his own definitions of *Vervollkommnung*. Darwin's uneven usage of "improvement" and "perfection" might even have suggested to Bronn that there was meant to be a distinction between practical improvement and morphological perfection. In any case, his German wording made that distinction more clearly than Darwin.

Darwin, Development, and Recapitulationism

One last set of issues that needs to be addressed has to do with embryonic development and recapitulation, the extent of Darwin's debts to the German transcendentalists on these subjects, and the sources of Haeckel's recapitulationism as well. Darwin's notebooks and early essays did indeed seem to endorse recapitulation of ancestral forms, but only weakly. His "Essay of 1844," for instance, began its section on embryology with an even-handed consideration of recapitulationism's pros and cons. On the one hand, Darwin wrote, "The unity of type in the great classes is shown . . . in the stages through which the embryo passes in coming to maturity,"[58] and he intimated that evidence from embryology could actually be used to make out evolutionary history. On the other hand, he did not claim specifically that the forms of adult ancestors were discernible in the embryo—or at least he avoided saying so in his own voice: "It has been asserted," he observed from a certain distance, "that the higher animal in each class passes through the state of a lower animal. . . ."[59] It is not clear what "the state" meant—probably something along the lines of Meckel's analysis of single organs or levels of complexity, rather than exact morphologies of ancestral species—but even this he did not let stand without qualification. Immediately he cited opposing scientific opinions, together with counterexamples of animals that, toward adulthood, regressed in complexity relative to their larval stages.[60] Clearly, Darwin was hedging his bets on the subject of recapitulation.

In his well-known analysis of Darwin's intellectual development, Dov Ospovat found that most of Darwin's efforts from 1838 through 1859 aimed to provide evolutionary explanations of all the laws and generalizations commonly recognized at the time in morphology, embryology, paleontology, taxonomy, and biogeography.[61] These efforts came to fruition in the later chapters of the *Origin* that demonstrated consilience. This need not mean, however, that Darwin accepted all the morphologists' laws and generalizations or that his theory depended on their validity. He was only preparing himself to argue that his theory was capable of explaining whatever the morphologists came up with.

Similarly, the main purpose of Darwin's discussion of embryology in the "Essay," as in the corresponding section of the *Origin*, was not to decide among competing conceptions of embryonic development, but only to show that no matter which one turned out to be most correct, the interplay of variation, heredity, and natural selection would suffice to explain it. The only firm commitment he made was to a rule of heredity that tended to preserve the relative timing of variations as they manifested themselves during development. He discussed it at first under domestication and verified it with breeders' experiences: "A much more important rule, which I think may be trusted, is that, at whatever period of life a peculiarity first appears, it tends to appear in the offspring at a corresponding age, though sometimes earlier. . . . I believe this rule to be of the highest importance in explaining the laws of embryology."[62]

Haeckel later repeated this generalization, called it a law, and built his own system of recapitulation upon it. But neither he nor Darwin ever said that favorable variations *necessarily* arose only at the end of embryonic or larval life. If they happened to do so, however, then natural selection left the earlier parts of the developmental process unchanged while accumulating changes to the later parts and the adult. Two varieties descended from the same parent stock, then, could inherit the original pattern of development up to the point when the distinguishing variations appeared. If those varieties continued to diverge, and new modifications continued to affect mainly the advanced embryo or young adult, then the developmental pattern would be conserved, even as distinct species and genera evolved. That would explain why similarities were often observed in the embryos of related species.[63]

Should the opinions of the morphologists have turned against recapitulation as the rule in embryology, that would have been no problem for Darwin's theory, because Darwin also had a ready explanation of the contrasting case, in which related species had dissimilar embryos. If new, favorable variations appeared early in the individual's development, then natural selection simply would not be able to preserve as much of the ancestral pattern of development as above. Darwin thought that such an outcome was most likely in animals with active and unprotected embryonic stages, like caterpillars and free-swimming larvae of marine animals, because they were subjected to strong selective pressures long before adulthood.[64] Haeckel later developed comparable ways of accommodating cases of nonrecapitulation.

On the whole, Darwin seemed to think the first scenario—conservation of the ancestral form in the embryo—would have been the more common one, but he did not make it the rule in the essay, and he was even more noncommittal in the *Origin*: "Hence, I conclude, that it is quite possible, that each of the many successive modifications, by which each species has acquired its present structure, may have supervened at a not very early period of life; and some direct evidence from our domestic animals supports this view. But in other cases it is quite possible that each successive modification, or most of them, may have appeared at an extremely early period."[65]

Either way, it fell far short of a full-blown theory of recapitulation. In Darwin's examples of variation late in development, some semblance of some ancestral embryonic form or forms was conserved in various descendant species, and Darwin took it as evidence that those species were related. He might also have been able to glean clues from the embryos as to the appearance of some ancestor, but Darwin gave no indication that he thought the sequence of embryonic stages would ever be readable as a historical record of modifications.

Only in one case did Darwin flirt with recapitulation of adult forms. This was when he speculated that, if variation were usually added on at the very end of development, the process would have become longer and more complicated with time. In the distant past, then, development might have been more direct, and embryos and adults might have looked more alike. In that case the recent embryo could indeed resemble both the ancestral embryo and the ancestral adult: "It may be argued with much probability that in the earliest and simplest condition of things the parent and the embryo must have resembled each other, and that the passage of any animal through embryonic states in its growth is entirely due to sub-sequent variations affecting *only* the more mature periods of life. If so, the embryos of the existing vertebrata will shadow forth the full-grown structure of some of those forms of this great class which existed at the earlier periods of the earth's history"[66] (emphasis in original).

Here Darwin came close to saying that some present-day embryonic stages resembled some adult ancestors, but only with a great deal of quali-fying and only with respect to the most distant ancestors within a class. There was still no talk of the embryo passing through an actual historical sequence of ancestral forms. It is hard to see how this made Darwin "a more thoroughgoing recapitulationist than his predecessors,"[67] especially in light of his concluding caveat of the chapter in the "Essay": "But I think our evidence is so exceedingly incomplete regarding the number and kinds of organisms which have existed during all, especially the earlier, periods of the earth's history, that I should put no stress on this accordance, even if it held truer than it probably does in our present state of knowledge."[68]

Darwin's stance only became more equivocal in the *Origin*: "It should also be borne in mind, that the supposed law of resemblance of ancient forms of life to the embryonic stages of recent forms, may be true, but yet, owing to the geological record not extending far enough back in time, may remain for a long period, or for ever, incapable of demonstration."[69]

In short, Darwin's treatment of recapitulation was more an attempt to explain it, whenever it could be shown to occur, than an endorsement of its use in working out evolutionary history. Other Darwinians, particularly Haeckel, made recapitulationism into an investigative tool and a research program.

Chance and Necessity in Early Responses to Darwin's Book

Darwin's efforts to portray natural selection as a *vera causa* and to satisfy the methodological standards of Herschel and Whewell did not win over those two philosophers or many physicists and physicalistically oriented biologists. Darwin's historical narratives were too many and too hypothetical, and the element of chance in them seemed to deny the apparent order in organic nature. Herschel, expecting to find a deterministic law in Darwin's *Origin*, instead of random variation and an unpredictable selector, is famously supposed to have disparaged natural selection as "the law of the higgledy-pigglety."[70] Evolutionists following transcendental or developmental models were equally disappointed in Darwin's unpredictable and purposeless mechanisms. They criticized natural selection for its lawlessness and indeterminacy, and they compared it unfavorably with the concept of *Entwicklung*.

The paleobotanist and pre-Darwinian evolutionist Franz Unger, for example, felt gratified and vindicated by the growing acceptance of species transformation and common descent, but he did not find in Darwin's approach quite the "physics of the plant organism," or of organisms in general, that he had envisioned. He found it too materialistic and haphazard to account adequately for man or for nature as an interacting, harmonious whole.[71] Ironically, Unger had had to defend his conception of evolutionary law against public attacks by a theologian in the 1850s. The theologian, Sebastian Brunner, had insisted on historical contingency in paleontology, on the grounds that no deterministic law should be able to dictate to the Creator what He should create or when. Now, in the 1860s, Darwin had done away with such laws and given nature some of the Creator's discretion.[72]

In his overview of the initial reactions to Darwin's theory, the comparative anatomist and histologist Albert von Kölliker counted the lack of law and plan as one of the gravest points against Darwinian evolution. The German embryologist Albert Wigand invoked the authority of Newton against Darwin's lawless approach, while the Swiss botanist Carl Nägeli took Darwin to task for trying to use random external influences to explain nature's order. He considered it quite unscientific to invoke chance as an explanation of anything, and he said it revealed Darwin's lack of qualification as a physiologist. As late as 1902, the botanist Julius Wiesner (Unger's successor in Vienna), continued to hold that random changes and environmental forces were unlikely explanations of progressive development in the plant and animal kingdoms. He missed the embryological analogy and a creative force, like the old *Bildungstrieb*, in Darwin's theory.[73]

Bronn stood out among German-language commentators for welcoming Darwin's emphases on historical contingency, adaptation, and the complex interactions among organisms and changing environments. His ambitions for an independent discipline of paleontology made him resist direct applications of embryological analogies to the history of life and left him open to alternatives such as Darwin's. Still, he did not accept Darwin's arguments without reservation, or interpret them exactly as would a British gentleman naturalist. He was at one and the same time Darwin's interpreter and critic, his ally in promoting a historical approach to life and his rival for priority and leadership in the dawning era of historical biology.

As an ally, he strove to make Darwinism palatable to himself and his fellow morphologists by treating natural selection, where possible, as a law of nature that acted with the same necessity and predictability as any law of physics. He also translated the *Origin* not simply into German, but into familiar morphologists' terms such as *Entwicklung* and *Vervollkommnung* (perfection). Avoiding neologisms, he used the older language in new and subversive ways to describe patterns of random variation, selection, and competitive improvement. The linguistic continuity by itself does not demonstrate either Bronn's confusion or the persistence in his writings of transcendentalist or developmental notions of type, progress, and purpose.

As a rival, Bronn took several opportunities to refer Darwin's readers to his own works as the more authoritative sources on the fossil record and the patterns or laws of progress that Darwin should have been using and explaining. He also tended to depict natural selection either as an extension of his own system, in which organic change was driven by the continual need for adaptation to changing environments, or as a flawed alternative to it. And whenever Darwin used his tactic of contrasting natural selection and common descent with "the accepted view" or issuing challenges to unnamed creationists to explain a puzzling phenomenon, Bronn construed it as a reference to his own successional theory and rejected the comparison.

Darwin continually underestimated Bronn as a rival and never thought of his work as the accepted view that needed to be challenged. He did not read Bronn's prizewinning essay until after writing the *Origin of Species*, and seems only to have mined the earlier *Geschichte der Natur* for factual information, mainly on variation and domestication, while ignoring the laws and explanations. Probably, he did what historians have continued to do to Bronn and took him for an old-fashioned German transcendentalist or developmentalist. That is why he did not think it "worth the labour," at first, to try to decipher Bronn's book review in early 1860.[74]

Bronn's Book Review

In his review of the *Origin*, Bronn responded directly and indirectly whenever he perceived that Darwin was overlooking or slighting his work. He set a somewhat derogatory tone at the very start by denying any great originality to Darwin. Instead, he positioned Darwin on the shoulders of two giants of geology, whose work he thought Darwin was following up, namely Charles Lyell and Bronn himself. First came Lyell. Darwin's book, Bronn wrote, was "A work, whose basic idea is liable to set the scientific world into even more motion than that once developed in the Lyellian *Principles*, which is continued here, in a certain way."[75]

The allusion was to Lyell's idea of uniformitarianism—using present-day processes to explain past events. In the case at hand, Darwin was using present-day variation and experience with artificial breeding to explain the history of varieties and species. But was the present-day process applicable in this case? Bronn found the application problematic because the present-day process had only been observed to produce new varieties, not new species. Thus the *Origin* might have been attempting to continue along the same lines as the *Principles*, but "whether with the same actual success [as Lyell], may be doubted, since there is no prospect at hand of mustering irrefutable evidence to the same degree as for the former work, although, of course, it seems just as impossible to provide decisive evidence to the contrary."[76]

Moreover, even if Lyell did not do so himself, variation had already been well studied, and the extrapolation from variation to speciation had been attempted before, and rejected, by Bronn: "Species can vary. This is generally recognized! Differences in food, habitat, climate, and some still-unknown causes bring forth varieties."[77] Here Bronn cited his own *Geschichte der Natur* in a footnote, as a prior and more authoritative source on the causes of variation. It was probably no coincidence that a brief report on Bronn's recent essay on island biogeography appeared on the same page of Bronn's journal, just above the review of the *Origin*.[78]

Bronn cited himself again toward the end of the review, and in between he was always weighing Darwin's theory against his own, or else considering it as a possible addendum to it. Bronn had already thought about variation, Malthusian overproduction, and struggle, and he had established that the maladapted went extinct. Thus the competitive and eliminative aspects of Darwin's theory did not interest him much in his review. What Bronn *was* looking for in Darwin, was the *creative* process that his own theory had left unspecified, and he wanted it described in terms of necessary laws and the forces that they governed. He did not find precisely that, but something close enough to be intriguing and to entice him into the translation project.

Bronn did not make all the connections Darwin could have wished for between artificial breeding and natural selection. He downplayed the selection process, focusing on the production of variant individuals rather than on the human agents that selected among them. Domestication and artificial breeding, in Bronn's estimation, only accelerated the known natural process of generating aberrant forms. They did not suggest any new kind of process: "Our cultivated plants and domestic animals teach us to what extent departures from the original type are possible even in a short time. When for every variation that is aimed for, man carefully selects those individuals that again deviate the most in the desired direction from the original type, he achieves in the relatively short time of a few dozen or a hundred years such extraordinary successes as could not come about by nature's method in ten- or a hundred times as long a period. The former case shows what might also be possible, with time, here in nature."[79]

Moving from the artificial to the natural, the key analogy, in Bronn's estimation, was in the fortuitous production of deviant individuals who could become the founders of new varieties and species, rather than the cumulative effects of selection: "If we find, however, that in this way, in hundreds or thousands of years, randomly appearing individual variations can turn into permanent races and these into species, then only hundreds of thousands of years are needed to further bring forth different genera from different species—and a few million years to bring forth different orders and classes from these. And since we have no shortage of time for this, there is nothing substantive left to raise as an objection to it, even if the explanations may indeed run into great trouble in matters of detail and especially in particular cases."[80]

Bronn was prepared to go a long way with Darwin. He agreed that environmental change and adaptation were the underlying causes of species turnover. He already defined species by community of descent from a number of founding individuals who might well have varied originally among themselves. The big change, from Bronn's point of view was that Darwin made those founding individuals the offspring of preexisting species, rather than separate creations or productions: "The means of progressive development, following Darwin, would of course be very different from the currently accepted view [i.e., Bronn's own], in that, in the continual drive toward adaptation to external conditions of life, the increasingly perfect and higher new specific- and generic-, etc. forms that appear would, in our view, be newly created; in Darwin's they would have originated [*entstanden*] out of the old."[81]

Bronn seemed quite willing to compromise with Darwin here, because, as the quote implies, improvements in adaptation could result in increasing perfection. He could make do without his separate laws of progress, if necessary. He was also willing to at least consider the possibility of species transformation. Indeed he had considered it before, but rejected it for lack of evidence that varieties could diverge limitlessly from the parent species. Nature's lawful means (*gesetzliche Mittel*) of returning species to the norm seemed to Bronn to preclude variation from accumulating in the Darwinian manner from the variety level to the species, genus, and above.

For Bronn, the crucial question became whether there were limits to variation and whether the process of variation could, in the long run, break the species-barrier. For if there could be common descent of species, then creative forces and successional theories of all kinds might have been rendered redundant, including Bronn's own: "Here there is only one of two possibilities: either his theory is incorrect (does not apply beyond the production of varieties), or, if it is correct, then variability is unlimited, i.e., there is no creation of the organic world, i.e., the force of nature has been found through which the organic world originated [*entstand*], and the assumption of a Creation is unnecessary."[82]

Bronn did not shrink from this prospect, but he had two major reservations. The lesser one concerned the continued lack of empirical support for the claim that there was no limit to the divergence among descendants of a single species. Darwin still had not provided the requisite support or uncovered any flaw in Bronn's reasoning about the countervailing lawlike processes. Indeed, Darwin had not taken any explicit notice of Bronn's arguments at all.

The greater of Bronn's reservations concerned the problem of ultimate origins. Citing recent experiments that seemed to rule out the spontaneous generation of simple organisms out of inanimate matter,[83] Bronn argued that Darwin had no way to account for the origin of the first species. Natural selection by itself could not be *the* sought-after force of species production, because some other force had to be behind that first creation. As long as the existence of this other force had to be admitted, what was gained by postulating natural selection in addition to it? Natural selection might be the redundant one.

Bronn's challenge to Darwin, therefore, was to account for the origin of all species, including the first: "Make organic matter with a cellular structure out of inorganic matter, proceed then to create seeds and eggs of lower organic species out of this organic matter—a task which modern science is equal to if it is at all possible. Then with the additional help of Darwin's theory, we can conceive of a natural force which might have brought forth all organic species. Thus, we will no longer be forced to seek recourse in personal acts of creation which fall outside the scope of natural law."[84]

Here Bronn cited his own *Entwickelungs-Gesetze* and his call for a new, naturalistic account of species origins. Such an account might yet be given, and once it was put forth, then Bronn would be willing to argue about the next problem, which was whether natural selection adequately accounted for adaptation, purpose, and complexity: "However, as long as this is not possible, the Darwinian theory remains all the more improbable, as it moves us no closer to the solution of the great problem of creation. And that is leaving entirely out of consideration how it is even thinkable that such an organism as a butterfly, a snake or a horse, etc., wisely calculated down to the last tiny fiber, could be merely the product of a blind natural force!"[85]

A Question of Lifestyle

If anything in the review gave Darwin the impression that Bronn was not understanding him properly, it was Bronn's treatment of natural selection. Bronn translated the term as *"Wahl der Lebensweise,"* which translates back literally as "choice of the way of life" or, more liberally, "lifestyle selection." According to Bronn, it was Darwin's main cause of variety production, and it worked as follows: Malthusian population growth led to competition and struggle for existence. Struggle forced many individuals to move to new places, try out new food sources, or somehow *select* a new means of making a living or a new environment in which to live. The individuals who selected a new lifestyle then become modified in consequence of their choices: "The reproduction of animals and plants is namely all too profuse to not always force one large portion of the offspring to choose new nourishment and altogether a new way of life than the other. This variant life-style demands and gradually also develops variant uses of the organs, variant faculties, and variant forms: if these same external causes persist from generation to generation, persistent races originate, which pass their variant characteristics on to their offspring, even under different conditions, so that one often cannot tell anymore whether one is looking at a species or a variety."[86] Darwin concluded from this rendering that Bronn was confusing natural selection with Lamarckian environmental effects and use-inheritance, as indeed he probably was, in a hasty first reading of the book.

To be as charitable to Bronn as possible, "lifestyle choice" made some sense in some contexts within the *Origin*. There were many passages in the chapter on natural selection where Darwin wrote about migration or about how the struggle for existence might force species to move into new "places in the economy of nature," almost as if they were making conscious choices. The section on "divergence of character" in particular was all about species occupying and adapting to new and diverse "places," and Darwin's metaphor of the wedges trying to displace one another from the face of nature also encouraged an interpretation of "places" as real estate. Bronn took Darwin to be saying that once species got to their new places, they were necessarily induced by the change in environment to vary and adapt, as also required by Bronn's own principle of adaptation, but on an individual rather than a species level.

That the changed lifestyle induced variation was not an unreasonable interpretation. Darwin sometimes did invoke the kind of Lamarckian environmental effects and use-inheritance that Bronn attributed to him, and Darwin was in general agreement with Bronn that a change in environment could destabilize heredity and bring forth aberrant individuals. This happened under domestication, according to both Darwin and Bronn. Bronn's reading accentuated what was familiar and acceptable to himself. Still, by failing to mention variation in competitive ability and differential survival in the quoted passage, Bronn implied that the Lamarckian causes of variation sufficed to create adaptations, without Darwin's notion of selection. That was the implication of *Wahl der Lebensweise* to which Darwin objected most.

One context in which Bronn's lifestyle selection was most difficult to reconcile with Darwin's idea of selection was in relation to domesticated plants and animals and the analogy between artificial and natural selection. Under domestication, the word "selection" clearly referred to something the breeder did, and not to any choice that the organism might have made. How could Bronn have glossed over this frequent usage of the term selection? Could Bronn have entirely missed the intended analogy between the artificial and the natural?

Darwin must have been asking himself the same question as he struggled to read Bronn's German and to make up his mind about the offer to translate the book. He soon wrote a very tactful letter to Bronn, in which he thanked him for taking on the translation project, corrected him on *Wahl der Lebensweise*, and enclosed some last revisions to the second English edition of the *Origin*, which was the one that Bronn translated. He also sent along a new "Historical Sketch" of contributions toward theories of selection or species transformation, to be inserted at the very beginning. Thanking Bronn was the easy part:

My Dear and Much Honoured Sir,

I thank you cordially for your extreme kindness in superintending the translation. I have mentioned this to some eminent scientific men, and they all agree that you have done a noble and generous service. . . . I thank you also much for the review, and for the kind manner in which you speak of me.[87]

Explaining the "Historical Sketch," however, was full of pitfalls to which Darwin seemed oblivious. The sketch glossed over most of German biology in a cavalier manner, and he was sending it unabashedly to an elder German biologist who could reasonably expect more credit than Darwin was giving: "I send with this letter some corrections and additions to M. Schweizerbart, and a short historical preface. I am not much acquainted with German authors, as I read German very slowly; therefore I do not know whether any Germans have advocated similar views with mine; if they have, would you do me the favour to insert a foot-note to the preface?"[88]

In fact, the "Historical Sketch" gave no credit to Bronn at all, while also undermining Darwin's excuse that he was unacquainted with the German literature. For Darwin did cite Bronn's *Entwickelungs-Gesetze*, but only as a source of information on Franz Unger's evolutionism. There was no indication that Darwin had found anything else of value in Bronn's books.[89] One might argue perhaps that Darwin rightly left Bronn out because, strictly speaking, he was not an evolutionist, but then, Darwin dealt with Richard Owen at considerable length, whose theory was no more evolutionary than Bronn's and whom Bronn considered an opponent. Most likely, Darwin really had read little or none of Bronn's 1858 book at the time. He wrote to Bronn in 1862 that he was finally reading the French version, and he expressed deep regret at not having gotten to it before writing the *Origin*.[90]

The omission is a further illustration of how Darwin brought his vaunted networking skills to bear almost exclusively on his British colleagues, to the detriment of his international reception. For his part, Bronn let the slight pass without comment, but also without taking up Darwin's suggestion that he add footnotes on German sources to the historical sketch. Bronn had better ways of inserting himself into the discussion—as a current authority, not as a forerunner or historical curiosity.

Finally, Darwin's letter turned to the problem of *Wahl der Lebensweise*. Darwin pretended he was only questioning Bronn's choice of words for "natural selection," and not his fundamental understanding of the mechanism. First he suggested that it was all his own fault for deliberately introducing an obscure term: "Several scientific men have thought the term 'Natural Selection' good, because its meaning is *not* obvious, and each man could not put on it his own interpretation, and because it at once connects variation under domestication and nature." He then tried to steer Bronn toward agricultural terminology and away from anything that might smack of Lamarckism:

Is there any analogous term used by German breeders of animals? "Adelung," ennobling, would perhaps be too metaphorical. It is folly in me, but I cannot help doubting whether "Wahl der Lebensweise" expresses my notion. It leaves the impression on my mind of the Lamarckian doctrine (which I reject) of habits of life being all-important. Man has altered, and thus improved the English racehorse by *selecting* successive fleeter individuals; and I believe, owing to the struggle for existence, that similar *slight* variations in a wild horse, *if advantageous to it*, would be *selected* or *preserved* by nature; hence Natural Selection. But I apologise for troubling you with these remarks on the importance of choosing good German terms for "Natural Selection."[91] (Emphasis in original)

Bronn took the hint. He did not use *Wahl der Lebensweise* in the translation, or even *Adelung*, which, as discussed below, was an odd suggestion for Darwin to make. Bronn switched to *Züchtung*, a general term for breeding or cultivation.

Bronn's first interpretation of natural selection as *Wahl der Lebensweise* appears to have been an overhasty pigeonholing of Darwin as a new variation on familiar themes from Lamarck or from Bronn's own work on the necessity of adaptation to environmental change. Malthusian overproduction combined with Darwinian struggle to shift groups of individuals to new environments or lifestyles, and either Lamarckian environmental effects or Bronnian laws of adaptation did the rest. Under the latter interpretation, it was not all that different from Bronn's system, except that the laws brought forth a succession of better-adapted individuals instead of species.

Darwin's letter cut off this line of interpretation and demanded that more weight be given to the selection process than to the production of the fitter variations, which no law of adaptation or progress was supposed to govern. Thus the letter called Bronn's attention to the crucial question that divided him and most German morphologists from Darwin, and this was not the question of species transformation per se, but of how to balance the rule of law and the unity of the sciences with chance and historical contingency. The next chapter will analyze Bronn's efforts to bridge this conceptual and methodological gap in the translation.

4

Bronn's *Origin*

On the whole, the Bronn translation never deserved its reputation for inaccuracy and distortion. There were some interpretive and linguistic problems, to be sure, but hardly more than is usual in any translation, and many of them have arisen only in the minds of modern readers who have ascribed anachronistic meanings to Bronn's words, and perhaps to Darwin's as well. Bronn's prose might have been somewhat disappointing to anyone who wanted him to wax as poetic as Darwin sometimes could, but it was certainly up to any reasonable standard of nineteenth-century scientific writing. Bronn tried faithfully to follow the original closely and to refrain from interjecting his own opinions overtly into the main text. He saved most of his objections for the end. The translator's footnotes were not very long or frequent, and they usually provided zoological or paleontological points of information or discussion of word choices. Bronn's voice intruded only in a few notes that referred the reader to his own work, suggesting a certain continuing irritation at not being cited, especially where Darwin was generous with credit to Owen or others. For example, in connection with the increasing prevalence of serially repeated organs or segments as one goes down the morphological scale, Bronn noted that "these and related questions were developed much more exhaustively" in his own *Gestaltungs-Gesetze* than they were by Owen.[1]

Rather than pick over the translation for every departure from an arbitrary ideal, I shall begin with just a sampling of individual infelicities and "errors," then move on to discuss the larger interpretive issues that arose when a German professor read the work of an English gentleman-naturalist and attempted to reproduce it in his own specialized language. Most of my examples are from Bronn's first edition (based on Darwin's second). His second edition (based on Darwin's third) ironed some of them out, but also introduced new wrinkles to go with Darwin's new material.

The problems began at the very beginning, with Darwin's title and subtitle: "The Origin of Species by Means of Natural Selection: Or the Preservation of Favoured Races in the Struggle for Life." Bronn had to chose among several possible German wordings for "origin," "natural selection," "favoured races," and "struggle for life," and, as discussed in greater detail below, his choices announced his point of view and his grasp of Darwin's metaphors. The problems multiplied in the first chapter, on artificial selection and domesticated animals, where the German professor displayed his unfamiliarity with the English gentleman's hunting dogs and fancy pigeons. Some of the force of Darwin's analogy between natural and artificial varieties got lost in translation, even when Bronn was able to find German names for the English breeds.

In addition to purely linguistic problems, there was the question of how delicately to treat the implications of the *Origin* for the human sciences and our self-understanding. As is well known, Darwin avoided human biology, venturing only a few short sentences on the subject in the last chapter, where he predicted: "In the distant future I see open fields for far more important researches. Psychology will be based on a new foundation, that of the necessary acquirement of each mental power and capacity by gradation."[2]

In the German, Bronn decided to change "psychology" to "physiology" (*Physiologie*).[3] It was probably not a simple oversight, excusable perhaps as a case of presbyopia, because he let it stand in the next edition. Yet it is difficult to explain what he was trying to convey with this change. If he wished to mask Darwin's suggestion that our mental powers (*Fähigkeiten des Geistes*) have evolved gradually, the change to "physiology" did not help much by itself, since the rest of the sentence conveyed Darwin's drift anyway. Perhaps he was trying to make a statement about the scope of physiology as a discipline that should include the mental faculties in its purview.

The most egregious liberty that Bronn took with the text, and which Francis Darwin complained about in the *Life and Letters*,[4] concerned Darwin's famous understatement in the same paragraph, that "Light will be thrown on the origin of man and his history."[5] Bronn omitted the entire line. It is generally assumed that Bronn considered the sentiment too radical to include in the book, but the matter is not so simple, because Bronn never showed any qualms about discussing human origins in biological terms before. He did so in his own *Geschichte der Natur*, even if he was still pious enough to have man created separately and embody morphological perfection (*Vervollkommnung*). Most puzzling is that Bronn also brought up the possibility of human evolution in his concluding critique at the end of the translation, not very far below the omission. There he wrote that humankind offered the most striking empirical evidence one could desire for Darwinian divergence of character among groups long isolated from each other: "What was the result of this isolation of individual human groups during such a long period of time? Just as many races as there are separate continents. . . ."[6] Was that any less radical a suggestion than Darwin's about the origin of man and his history? Perhaps Bronn's objection was methodological, on the grounds that Darwin was not throwing light on human biology, but drawing support from it.

The most amusing mistranslation, and one that is easy to miss, is from the opening lines of Darwin's third edition:

When on board H.M.S. *Beagle* as naturalist, I was much struck with certain facts in the distribution of the organic beings inhabiting South America, and in the geological relations of the present to the past inhabitants of that continent. These facts, *as will be seen in the latter chapters of this volume*, seemed to throw some light on the origin of species.[7] (Emphasis added)

[Diese Thatsachen schienen mir, *wie sich aus dem letzten Kapitel dieses Bandes ergeben wird*, einiges Licht über die Entstehung der Arten zu verbreiten]

In the second edition of the translation, the unspecified and plural "latter chapters," where all the light was, became "*the* last chapter of this volume" (emphasis added).[8] A Freudian slip, perhaps? For if one flipped to the back of the German volume, one found not Darwin's chapter 14, "Recapitulation and Conclusion," but a supernumerary chapter 15, containing Bronn's commentary. It was not an appendix or postscript, but a regular, numbered chapter, albeit with the title "Schlusswort des Übersetzers" (Translator's conclusion) indicating its authorship. Bronn's chapter appeared in this edition unchanged from the first, but it was deleted from the third German edition, with Darwin's hearty approval, by the zoologist J. Victor Carus, who took over as Darwin's principal German translator after Bronn's death.[9]

Bronn's Critique: The Pros

Bronn never claimed to be a passive conveyor of Darwin's ideas. He viewed himself quite justifiably as by far a better-established authority than Darwin on the history of the organic world, and he felt he had not only the right but the obligation to introduce Darwin's theory to his German audience in his own way. He could also cite Darwin himself for the suggestion and the license to apply his judgment and append his editorial comments to the translation.

In these comments, Bronn began by addressing the reader familiarly in the second person, and asking rhetorically about your state of mind after reading this wonderful book: "You wonder how many of your previous views about the most important natural phenomena it has left untouched, and how many of your previously firm convictions still stand firm?"[10] Bronn offered graciously to help the reader sort things out: "Since we have undertaken, in accordance with Mr. Darwin's wishes, to translate his work into German, we believe we owe the reader an account of our own previous views on many of the individual questions explored by the author, and our view of his theory as a whole, as well as the influence that theory has had on our own manner of thinking."[11]

Bronn's tone was friendly, even admiring. He seemed to identify personally with Darwin, whom he described in glowing terms with which he might have wished to be described himself: "a respected naturalist," developing a new approach to the study of life and its history, and "viewing in an original and incisive manner, all the facts that he has been collecting and observing for twenty years and over which he has constantly been brooding and pondering for twenty years."[12] He ascribed to Darwin the very goal that he had set himself but failed to reach, namely, discovering the necessary, fundamental law of species creation.

As Bronn put it, Darwin sought:

To solve the previously seemingly insoluble problem, the greatest mystery of natural science, and to establish *one* idea, *one* fundamental law of becoming and being in the organism-world, which would govern that world in time and space just as gravitation prevails in the heavenly bodies and elective affinities in all of matter, and to which all other laws that have been proposed for the organism-world would be reducible. It is the law of development by means of natural selection [*das Entwickelungs-Gesetz durch Natürliche Züchtung*] that rules in all of organic nature as well as in the natural system and in the individual through time and space.[13] (Emphasis in original)

Bronn was evidently considering natural selection as a potential contribution to his own system and presenting it as an appropriately *wissenschaftlich* device, a developmental law, embedded in a hierarchy of physical and chemical laws, which unified and explained the phenomena.

After introducing Darwin as a naturalist, Bronn proceeded to give a brief summary of his own theory, presenting it as the established view against which Darwin's needed to be weighed, or upon which Darwin could have been expected to build. Bronn's presentation suggested just what gap remained in the established view to be filled by Darwin's new law of natural selection. Bronn was still searching for a naturalistic explanation of species creation, preferably in the form of a physicalistic law or force: "Therefore, for lack of any other species-forming force of nature [*Natur-Kraft*] . . . [we] have found it necessary, for the time being, still to invoke a Creation, however, with the explicit note that such an assumption of personal activity on the part of the Creator stands in contradiction to the way the rest of Nature is governed."[14] Bronn had never detected such a creative force, and he was quite unsatisfied with the theistic alternative. The rest of the critique judged whether natural selection was a fit substitute for such a force.

Bronn's summary of Darwin's theory began not with artificial selection or the mechanism of variation, struggle, and selection—the way Darwin's book began—but with an outline of the big picture of organic history, Darwin's narrative of the Creator breathing life into one or just a few original forms, and these forms giving rise evolutionarily to all the rest. Variation, struggle, and selection came only later, and Bronn never said much about artificial breeding at all. He did not seem to think the analogy to artificial selection was important. He delved much more deeply into what he considered the crucial assumptions behind Darwin's mechanism of change: first and foremost, Darwin required limitless variation, continual adaptation to local conditions, and the splitting and limitless divergence of lineages.

Bronn's reservations became apparent as he reiterated Darwin's claim that there was no limit to the cumulative effects of selection or to the divergence of a variety from the species norm. Could a variation give rise to a variety, a variety to a species, and so on up the taxonomic hierarchy? Before considering whether the claim was true, he asked whether there was any necessary law against it, and he conceded that there was none: "There is no natural cause and no logical reason to assume that the amount of gradual change finds a limit anywhere."[15] That meant the matter was at least open to empirical investigation, and that his own conclusions from *Geschichte der Natur* might have been subject to revision, if Darwin made his case convincingly.

Bronn noted approvingly that Darwin always related the utility of a variation to the environment in which it occurred. Different environments necessitated different adaptations, and one would therefore have expected the process of adaptation to require continual adjustment and readjustment to environmental change. Here he could see that Darwin agreed with him on the principle of adaptation, even if they differed on evolutionary continuity. He also noted with approval that Darwin's conception of the environment included not only its physical features but also the complex interactions among organisms, again a fundamental claim of Bronn's. Then he paraphrased natural selection in his own terms and made it seem to provide *necessarily* for perfection and diversity at the same time as adaptation: "Continual and enduring multiplication and dispersal of the perfected [*Vervollkommneten*] victors and continual 'extinguishing' of the—because of lesser perfection—vanquished. The more life forms originate, the more diverse [*manchfaltiger*] therewith again become the conditions of life. Therefore also continual change, perfection, and diversification of a portion of the life forms (although others disappear), and not by chance, but as a necessary, lawful phenomenon!"[16]

Bronn even saw the way to explain and unify his many laws of progress and reduce them to effects of natural selection. He only had to assume that progress, in the various ways he had defined it, tended to be adaptive. Here he specifically mentioned the development and "perfection" of individual organs by use-inheritance, the differentiation of serially repeated organs, the diversification of forms, and the conservation of early stages of embryonic development. These were all measures of progress that Bronn had identified in previous works and that Darwin could now explain as effects of natural selection.

. . . And the Cons

Just when it seemed that Bronn was about to appropriate Darwin's theory with enthusiasm, he turned to the other side of the coin and came up with two major objections and a list of many lesser problems. The lesser problems were ones that stemmed from the empirical shortcomings of Darwin's case, and Bronn was willing to put them aside until they could be solved during the course of further research. The lack of proof that there was no limit to divergence from the species norm fell into this category, as did the permanence and heritability of whatever divergence did occur. Bronn recalled the many lawlike means by which nature kept species coherent, and he noted that Darwin had not countered them all with empirical examples. Among other of the lesser problems, Bronn was also dissatisfied with Darwin's explanation of why some of the most primitive organisms lived on to the present day, unimproved by natural selection. And he demanded more and better analysis of what it meant for one variety to be superior to another. He found Darwin vague and evasive when it came to defining what we would now call "fitness," and explaining how it related to progress and perfection.[17]

More seriously, Bronn rejected Darwin's characteristic method of providing many separate, hypothetical historical narratives to explain evolutionary outcomes and letting "improved" or "favoured" refer to different kinds of qualities in different situations. Accepting such a strategy would have defeated Bronn's efforts to place paleontology on a *wissenschaftlich* footing by deriving and quantifying its general laws of change and progress. Darwin's approach almost evoked the image of the prescientific natural historian, who only collected and described individual cases, without systematizing and illuminating them properly. Still, Bronn had no definite grounds for rejecting all those historical hypotheses.

The two major problems were also matters of principle—violations of Bronn's ideal of *Wissenschaft*, and explanation in terms of law and necessity—and Bronn was not prepared to compromise on these at all. Above all, Bronn objected to the infinite graduality and multidirectionality of variation. He called it his *"first and most serious objection against the new theory*, because it touches its foundations"[18] (emphasis in original). In contrast to the orderly and directional variation that would have been produced by Lamarckian environmental effects or the actions of a *Bildungstrieb* or creative force, Darwinian variation went everywhere without constraint, yet somehow managed to yield discrete taxonomic groups, rather than infinite, unclassifiable intergradations between species: "The developing varieties, according to Darwin, are not as a rule formed by external influences and never as a product of a special, internal formative force that persists in varying in a certain direction, but rather by imperceptibly small deviations that run quite randomly in all possible directions, of which those that are useful to the organism have the best prospects for survival. . . . Thus the varieties will not separate themselves as such neatly and completely . . . from the parental form."[19] The burden was therefore on Darwin to explain why the whole orderly system of nature that one observed resulted *necessarily* from the process of natural selection. A plethora of hypothetical narratives, each telling how one coherent grouping or another could have been formed, would not do.

Bronn's *"second essential objection"* (emphasis in original) concerned the justifiability of rejecting a Creator or creative force in favor of species transformation.[20] Bronn faulted Darwin for his evasive and unrealistic account of the origin of the first species. Since there was no empirical evidence for spontaneous generation, any transformationist had to rely on unknown creative forces for those first species. Bronn would not let

Darwin apply a double standard and criticize the successionist for using the same sort of unknown creative forces. The problem was the same, even if the successionist had to invoke the forces more often than the transformationist, and Darwin had not solved it or eliminated creative forces from biology.

All told, Bronn was of two distinct minds. On the one hand were these objections that weighed heavily against Darwin: "We have brought forward several objections against it, and our personal ability to adopt it as it stands is even less than these objections would suggest." On the other hand, Bronn could see no naturalistic alternative to natural selection. In the very next sentence, he wrote, "But it leads us onto the only possible path! It is perhaps the fertilized egg, out of which the truth will slowly develop; it is perhaps the pupa, out of which the long-sought law of nature will emerge. . . . Or perhaps we already have the law we sought before our eyes, but see it only through a kaleidoscope."[21]

Bronn did see one *wissenschaftlich* virtue in Darwin's theory that might possibly have tipped the balance Darwin's way. It did not quite convince Bronn, but it led him to predict that the future would indeed belong to the Darwinians. That virtue was the theory's ability to provide a unifying causal explanation of all biological phenomena. Darwin's efforts to demonstrate Whewellian consilience were not wasted on Bronn:

The possibility, under this theory, to connect all the phenomena in organic nature through a single idea, to view them from a single point of view, to derive them from a single cause, to take a lot of facts that previously stood separately and to connect them most intimately to the rest and show them to be necessary complements to those same facts, to strikingly explain away most problems without proving impossible with respect to the remaining ones, gives this theory a stamp of truth and justifies the expectation that the great difficulties that remain for this theory will be overcome at last.[22]

In the end, Bronn recognized that despite all the problems he saw in Darwin's theory at that moment, it might well have had an illustrious future ahead of it, and he issued the cautious prediction that "The Darwinian theory will probably never quite go under!" and that some naturalists and even more nonnaturalists would fall under its spell even before all of its problems were addressed. For his part, Bronn said he could neither go back to divine Creation nor embrace Darwin's new theory, but would continue to hold to his published views until something clearly superior came along: "For it seems to us more forthright and consistent, at least, to persist in the old scientifically untenable standpoint, with the expectation that in the wake of the clash of opinions, a tenable theory will develop, become clear, and ripen. . . . Only out of the clash of opinions will the truth emerge, and the originator of this theory [i.e., Darwin] will himself no doubt experience the great gratification of having opened a new path for scientific research."[23]

Darwin expressed satisfaction with the German translation and accepted the criticism graciously, even before actually reading much of it. He wrote to Bronn:

On my return home, after an absence of some time, I found the translation of the third part of the 'Origin,' and I have been delighted to see a final chapter of criticisms by yourself. I have read the first few paragraphs and final paragraph, and am perfectly contented, indeed more than contented, with the generous and candid spirit with which you have considered my views. You speak with too much praise of my work. I shall, of course, carefully read the whole chapter; but though I can read descriptive books like Gaertner's pretty easily, when any reasoning comes in, I find German excessively difficult to understand. At some *future* time I should very much like to hear how my book has been received in Germany. . . .

I shall ever consider myself deeply indebted to you for the immense service and honour which you have conferred on me in making the excellent translation of my book.[24]

It took several months for Darwin to bite the bullet and work through Bronn's German, but he still found nothing very objectionable and much to like in Bronn's concluding words about scientific truth emerging out of their clash of opinions: "I ought to apologise for troubling you; but I have at last carefully read your excellent criticisms on my book.—I agree with much of them, & wholly with your final sentence. The objections & difficulties, which may be urged against my view, are indeed heavy enough almost to break my back; but it is not yet broken!"[25]

Interpretive Issues: The Artificial and the Natural

All together, Darwin's *Origin*, Bronn's review, translation, and critique, and the few surviving letters between the two naturalists raise a number of important scientific and historical questions about Darwin's argument and its cogency for readers who did not share many of Darwin's biological and cultural assumptions and experiences. I have organized the problems into three overlapping areas: how to understand and weigh the importance of the analogy to artificial selection; how to relate Darwin's conception of progress through competition to other kinds of improvement and perfection; and how to deal with the problem of ultimate origins and the apparent order and purpose in nature.

The analogy between artificial and natural selection played a number of important roles in Darwin's argument. It made the case for the transformative power of selection as a process with cumulative effects over many generations, and it supplied some of Darwin's most detailed examples of modification and divergence of lineages. Darwin's account of the fancy pigeons was especially important as a potential answer to Bronn's objections, since it featured extensive and persistent divergence of varieties, the disappearance of intermediate forms, and the retention of sufficiently many ancestral characteristics to provide evidence of common descent.

But Darwin's analogy to artificial selection was designed to play most effectively to British audiences. Its methodological and rhetorical roles in the overall argument did not translate well into German. Methodologically, it was supposed to establish natural selection as a *vera causa* in the sense of Herschel, but that was not an issue for Bronn. For Bronn the legitimacy of natural selection as a causal agent was founded upon the logical and natural necessity that the maladapted not survive. Empirical support for its existence and efficacy could have been sought directly in nature, in organism-environment relationships, the way that Bronn had supported his own principle of adaptation. No analogy to human artifice seemed called for. If anything, the analogy undermined the idea that a necessary, deterministic law was involved, since a breeder could just as easily select capriciously or by unpredictable criteria. As a uniformitarian, Bronn might have been interested in artificial selection as a present-day process to be invoked in explanations of the past, but, again, Bronn would not have seen the need to use breeding practices for this purpose. The relevant present-day processes, for him, would not be human activities at all.

Rhetorically, Darwin's use of artificial selection echoed Paley's much-admired *Natural Theology*, which likewise began with a compelling appeal to human artifice as an analogy to what a greater power could accomplish. But any appeal to Paleyan logic was lost on the German professor, who was trained in a tradition of biological research that had made its break from theology long before Cambridge. He learned his natural history in a philosophical faculty, quite separate from the theological. Although Bronn made use of natural theology in *Geschichte der Natur* in the 1840s, where he inferred the existence of a plan of nature and ascribed species production to divine intervention, he had been distancing himself from it ever since. And even there, he had not made the

Creator into an anthropomorphic Technician, but made Him behave much more like a predictable force of nature.

Perhaps the most important rhetorical function of the chapter on artificial selection was to let Darwin begin his long argument and introduce his new concepts with reference to familiar, domestic animals. Once the reader saw the pigeons and hunting dogs in Darwin's light, he or she could follow Darwin more willingly into the more exotic territory of the Galapagos Islands or the entangled banks of the tropics. But this rhetorical tactic presupposed a certain familiarity on the reader's part with British hunting dogs, fancy pigeons, and farm animals. How many such animals was a German university professor likely ever to have studied or even seen? Bronn barely mentioned artificial selection in his critical chapter, and I take this as a sign that he did not think it was very important. He appeared to miss or discount every role of Darwin's analogy. Darwin's domestic examples simply did not speak to Bronn, and there is reason to believe that he did not really know what domestic varieties Darwin was referring to, or, if he did, that he could not picture them and their diverging characteristics.

Bronn's knowledge of zoology was encyclopedic, and no one was better qualified than he to translate all the zoological terminology in the later chapters of the *Origin*, but the opening chapter on domesticated plants and animals gave him unexpected trouble. Even though he had made a detailed study of plant and animal breeding, with particular attention to British achievements, Bronn was not completely familiar with all of Darwin's examples—or at least not familiar with them in the same way as Darwin. In his *Geschichte der Natur* he had counted varieties and discussed rates of production and what conditions best facilitated the breeder's work. He was not much concerned with documenting the precise morphological or behavioral changes that the breeders had achieved. He relied on published sources and most likely had not actually seen specimens of many of the breeds he discussed. Most of his readers, both inside and outside the academic community, probably knew even less about them. Bronn had to seek expert help to find German names for Darwin's examples, and still he was left with uncertainties and ambiguities. Darwin's references to "laughers" and "pouters" and "tumblers" and "runts"—all breeds of fancy pigeon—gave him almost as many headaches as the British hunting dogs. These examples were not incidental, but central to Darwin's argument, and Darwin wrote as if they were all commonly known, as indeed they would have been to British gentlemen.

Consider Darwin's simple statement that "The trumpeter and laugher, as their names express, utter a very different coo from the other breeds."[26] In German, *der Trompeter* seemed like the obvious equivalent for the former breed, but that turned out to be a false cognate. For "laugher" Bronn would have liked to use the philogically similar *Lachtaube* (laugh-pigeon), but zoologically the German word referred to the ringneck dove, a completely different species from all the others in Darwin's exposition (*Columba risoria* instead of *C. livia*), as Bronn had to explain in a footnote.

What other German breed could have fit here? It would have had to be named for the funny sound of its coo. Bronn settled on *die Trommeltaube*, literally the drum pigeon, with a coo that was likened to a drumbeat. However, that probably was still not what Darwin meant, because in a later discussion of pigeon breeding (in *Variation under Domestication*), Darwin gave *Trommeltaube* as the German for the trumpeter rather than the laugher. (Darwin provided no German equivalent for the laugher.)[27] In other cases, Bronn gave up and left variety names untranslated, for example the runt, but that can also be confusing, especially if the reader knows a little English, because the runt was the largest of the fancy breeds.

Bronn wrote to Darwin for help with the British hunting dogs, and he repeated Darwin's answer in a footnote, which I think confused matters further. At least it confuses me. Here is my translation of Bronn's translation of Darwin's helpful explanations:

Mr. Darwin provides me with the following information about the English races of dogs that are mentioned here:

The *Jagdhund* (spaniel) is small, coarse-haired, with droopy ears, and barks when on the track of game;

The *Spürhund* (setter) is likewise coarse-haired, but large, and when it picks up the scent of game, it presses itself, without a sound, against the ground [on the track??—Bronn's bracketed insertion], motionless for a long time;

Finally, the *Vorstehehund* (pointer), corresponds to the German *Hühnerhund* [bird-dog], and is in England large and smooth-haired.[28]

[Herr Darwin erteilt mir über die hier genannten Englischen Hunde-Rassen folgende Auskunft:

der Jagdhund (Spaniel) ist klein, rauhaarig, mit hängenden Ohren und gibt auf der Fährte des Wildes Laut;

der Spürhund (Setter) ist ebenfalls rauhaarig, aber groß, und drückt sich, wenn er Wind vom Wilde hat, ohne Laut zu geben lange Zeit regungslos auf den Boden [auf die Fährte??];

der Vorstehehund (Pointer) endlich entspricht dem Deutschen Hühnerhund und ist in England gross und glatthaarig.]

Thus, there was uncertainty not only about the names of the breeds, but about precisely how they look. The German words, even when they were technically correct, did not always convey the intended picture, especially in the case of the pointer, which got two German names plus a verbal description, before the reader was asked to accept it as an illustration of Darwin's claim: that it has diverged, by means of unconscious selection, from, of all things, a *Spanish* breed:

It is known that the English pointer has been greatly changed within the last century, and in this case the change has, it is believed, been chiefly effected by crosses with the fox-hound; but what concerns us is, that the change has been effected unconsciously and gradually, and yet so effectually, that, though the old Spanish pointer certainly came from Spain, Mr. Borrow has not seen, as I am informed by him, any native dog in Spain like our pointer.[29]

[Es ist bekannt, daß der Vorstehehund im letzten Jahrhundert große Umänderung erfahren hat, und hier glaubt man seye die Umänderung hauptsächlich durch Kreuzung mit dem Fuchs-Hunde bewirkt worden; aber was uns berührt, das ist, daß diese Umänderung unbewußter und langsamer Weise geschehen und dennoch so beträchtlich ist, daß, obwohl der alte Vorstehehund gewiss aus Spanien gekommen, Herr Borrow mich doch versichert hat, in ganz Spanien keine einheimische Hunde-Rasse gesehen zu haben, die unserem Vorstehehund gliche.]

The comparison in Bronn's footnote of the English pointer to German varieties turned out not to be very relevant and even distracted from the comparison Darwin wished to make.

In the case of the pigeons, too, the unfamiliarity and inaccessibility of Darwin's examples obscured the subtle argument that Darwin was making. He was trying to strike a delicate balance in the readers' minds, convincing them that the domestic varieties were so radically different from one another as to suggest divergence into distinct species or even genera, yet so similar as to be recognizable as descendants of a common wild ancestor. "*Compare* the English carrier and the short-faced tumbler," Darwin wrote, assuming that readers could picture these pigeons, "and *see* the wonderful difference in their beaks, entailing corresponding differences in their skulls"[30] (emphasis added). But the world of the pigeon fanciers was foreign to Bronn and probably to most of his academic readers, and they could not see or bring to mind the comparisons that Darwin wanted them to make. And even if they had access to pigeon fanciers and their stocks, it was not always clear which breeds Darwin wanted them to look at or whether they looked the same in Germany as in England. It is not that Bronn and German academics would have been unable to follow the logic of this argument—quite the contrary. But the examples did not have the intended effect. They had to be explained and footnoted and qualified so much that Darwin's evidence lost the vividness it would have had for British readers.

Selection and Perfection

Translating the word "selection" and the key phrases "natural selection" and "sexual selection" was also a difficult matter, considering the many allusions and connotations of Darwin's terms. In English, the desired analogies among the artificial, the natural, and the sexual were underscored by the common appearance of the word "selection" in the names of all three processes. The wording also emphasized that the analogies were not to any and all of the breeder's or gardener's techniques, such as grafting, transplanting, hybridizing, or changing foods and environments, but specifically to the practice of choosing among the available variants for further breeding. Also, the word worked well with Darwin's metaphorical depiction of natural selection as an anthropomorphic but superhuman agency, "daily and hourly scrutinizing" all variation, and making intelligent and benevolent decisions like a Paleyan Designer.

No translation could reasonably be expected to preserve every one of the above connotations. Bronn had to choose a word that would at least convey the meanings of "selection" that he considered the most important. Bronn's priorities, however, were not the same as Darwin's. He was less interested in pushing the analogy to artificial selection and had no desire to see natural selection personified, or to flirt with natural theology. In order for Bronn to take natural selection seriously and portray it in as positive a light as he could for his German colleagues, he had to treat it as a regular and predictable law of nature or as a necessary relationship between organisms and environments. Such considerations pointed him toward formulations that did not connote any agency in nature that might have done the selecting, like a breeder.

Recall that in response to Bronn's unfortunate first approximation, *Wahl der Lebensweise,* Darwin had suggested *Adelung*—"ennobling" or "refinement"—as a possible, but "perhaps . . . too metaphorical"[31] German breeder's term to use instead. Darwin probably meant (or should have meant) *Veredelung*, a related word that likewise could be translated as ennobling, but had the additional agricultural meaning that Darwin was grasping for and that *Adelung* lacked. *Veredelung* referred either to the specific practice of grafting, or, generally, to the improvement of a stock.[32] But this, too, was a poor choice for Darwin's and perhaps also for Bronn's purposes. It failed to single out the breeder's technique of selection, and the association with grafting suggested instead the perpetuation of vegetative changes in more of a Lamarckian manner. Taken to mean "ennobling" rather than "grafting," the word might seem to imply improvement on some other scale of refinement or nobility instead of utility to the breeder.

Bronn chose the terms *Züchtung* for the human art of breeding or cultivation and *natürliche Züchtung* for natural selection. This choice had the advantage of alluding to agricultural practice, but it still did not directly indicate that *selection* was the relevant one of the breeder's methods. Bronn compensated somewhat for this shortcoming by introducing the term in connection with Darwin's first mention of artificial selection, along with the further specification that it meant *"Auswahl zur Nachzucht"* (choice [of individuals] for subsequent breeding). Bronn acknowledged in a footnote that *Züchtung* by itself could be interpreted to include breeders' techniques other than selection, and that *Auswahl zur Züchtung* (choice for breeding) would therefore have been more precise. He also said that he had considered the neologism *Zuchtwahl*, literally "breeding-choice," but decided for unstated reasons that it would best be used for sexual, not natural selection.[33] Hence, he arrived at the poorly matched coinages, *sexuelle Zuchtwahl* (but sometimes also *geschlechtliche Auswahl* or even *sexuelle Züchtung*) for sexual selection, but *natürliche Züchtung* for natural selection. As a result, when Darwin wrote about "artificial," "natural," and "sexual selection," it sounded consistently like three versions of the same process, but it was not so in Bronn's German. In later German editions, Carus switched to *Zuchtwahl* for selection of all kinds, and the word is still commonly used. Another possibility would have been *natürliche Auslese* (natural harvesting or picking), which is also in use today, as is the borrowed word *Selektion*.

Bronn's problem with *Zuchtwahl*, I believe, was the implication that where there was a *Wahl* there was also a *Wähler*, an agent to make the choice. Especially after his exchange with Darwin over *Wahl der Lebensweise* in the book review, he must have been wary of ascribing agency to plants and animals. He was also trying to portray natural selection as an impersonal, necessary law of nature, rather than a decision-making agent, an effort that Darwin's metaphorical language continually frustrated. He was willing to acknowledge that individual organisms were free to choose mates in the course of sexual selection, so that there could be a *Wahl* in *sexuelle Zuchtwahl*; but there were no choices to be made under natural selection, hence *natürliche Züchtung*. *Züchtung* still linked natural selection to breeders' practices, in accordance with Darwin's intentions, but it also expressed Bronn's caveat about the element of choice.

Further interpretive and linguistic problems revolved around the question of what, exactly, natural selection selected, favored, or improved. Bronn complained in his critical chapter that Darwin avoided specifying what qualities gave one form an advantage over another in the struggle for life, or in what "improvement" generally consisted. In his own work, Bronn took great care to analyze his conception of progress and perfection and to set out criteria for measuring it. He now expected Darwin to be equally precise, and if possible quantitative, in his analysis of competitive improvement (Darwin rarely used the term "fitness" in early editions of the *Origin*). Bronn clearly wished to relate competitive improvement to progress and perfection.

In the main text, Darwin asserted, to Bronn's chagrin, that other naturalists had so far failed to define satisfactorily what they meant by "higher" and "lower" on the scale of nature. Having always laid out his criteria for progress in great detail, Bronn protested in a footnote. He contradicted Darwin with an example of how to rank two faunas relative to one another, based on the proportions of species in each that belong to higher and lower taxonomic groups (higher and lower groups being identified by Bronn's various morphological criteria).[34] Meanwhile, in the main text, Darwin went on to say that the only reasonable scale of nature was a ranking by competitive superiority. Observing, for example, that many British species had colonized New Zealand and replaced native forms, Darwin inferred that the British fauna contained the better competitors and was higher in that sense.

Bronn let the matter rest in the main text, but retaliated in his concluding chapter, where he referred back to these provocative passages and demanded to know precisely what made the British species better competitors. If anyone's criteria of highness or lowness were poorly laid out and analyzed, Darwin's were: "If Darwin cites the empirical observation that a large portion of the *British* flora, compared to that of *New Zealand*, is so perfected [*vervollkommnet*] that it replaces it, then we would have hoped, even if only in individual cases, to see it shown wherein this superiority [*Überlegenheit*] lies"[35] (emphasis in original). Bronn capitalized on Darwin's vagueness by suggesting an equivalence, or at least an overlap, of morphological perfection with competitive superiority. He did so through his frequent usage of the word *vervollkommnet*—both here and throughout the text—for "improved" as well as for related descriptors of directions of change.

Bronn's usage of *Vervollkommnung* appeared most prominently in the subtitle of the book. There, Bronn rendered Darwin's "preservation of *favoured* races in the struggle for life" as "*Erhaltung der vervollkommneten Rassen*," that is, preservation of the *perfected* races. That certainly raises red flags for modern readers, who associate "perfection" with pre-Darwinian concepts of progress on linear scales of nature. In later editions, Carus changed *vervollkommneten* to *begünstigten*, which was a more literal rendering, but which still had disadvantages, at least from Bronn's point of view. Like Darwin's original "favoured," the word might have seemed to imply the existence of an external agent to display the favoritism, something that Bronn would not want to accept. It also implied a value judgment and perhaps a scale of favorability, but failed (as did the original English) to convey what sorts of things natural selection was supposed to favor. Bronn probably thought he was improving the subtitle by specifying that the (not only morphologically, but by his reasoning, also competitively) more perfected races were the ones that survive.

Bronn did not use *Vervollkommnung* strictly for morphological perfection. He also used it often where Darwin wrote "improvement," even in contexts where it was clear that *competitive* improvement was meant. For example, in Darwin's "Recapitulation and Conclusion," Darwin listed among his "laws" of evolution: "A Ratio of Increase so high as to lead to a Struggle for Life, and as a consequence to Natural Selection, entailing Divergence of Character and the Extinction of less-improved forms."[36] In translation, "less-improved" became "*minder vervollkommnet.*"

Curiously, Bronn also did not always use *Vervollkommnung* for "improvement." He consistently avoided suggesting such an equivalence in the context of artificial selection, where he substituted the agricultural term *Veredelung*. Through this verbal differentiation, Bronn expressed his low opinion of Darwin's analogy between the natural and the artificial process, and implied that the selector's actions did not necessarily lead to the same kind of perfection to which nature tends. As discussed in the previous chapter, Darwin had left himself open to this sort of interpretation, because he himself did not always use the same words for artificial and natural improvements. Bronn took the opportunity to strengthen the contrast.

Progress and Perfection in Darwinism

Historians have been too quick to link Bronn to Lamarckism, *Naturphilosophie*, and transcendental morphology, based on his usage of the term *Vervollkommnung*. For Junker, that one word pinpoints the fundamental difference between Bronn and Darwin, between a view of nature that had it progressing toward some idealized goal of perfection, and one that let progressive change be driven by nothing but selective advantages.[37] Nyhart may be more nearly correct when she says that to Bronn "favoured races" could not have meant mere competitive superiority, but Bronn gave more weight to competitive superiority than she allows.[38]

Such interpretations read too much of German idealistic morphology into Bronn. Progress toward greater perfection was only one of Bronn's laws of organic change; adaptation and survival also came into play in his system. Not only that, but Bronn measured progress by concrete criteria, such as increasing division of labor, reduction in the number of serially repeated parts, and increasing protection of important external organs—directions of change that would often be good for the organism in practical ways. These were no longer abstract, idealized scales of perfection, but of qualities that could contribute to survival.

Bronn's usage of *Vervollkommnung* was his means of linking Darwinian fitness to his own conception of progress. The word represented a tentative move by Bronn toward a synthesis of perfection with selection, under the assumption that morphological progress came about because the "favoured races" were simultaneously the higher on Bronn's morphological scales. This synthesis, if successful, would have fulfilled Bronn's longstanding goal, expressed in his *Geschichte der Natur*, of eventually reducing all of his other laws to side effects of the Fundamental Law of Adaptation. The subtitle of the German *Origin* was perfect for announcing this as Darwin's potential contribution to Bronn's theory. It also provided a form of insurance against the possible success of Darwin's theory. Under Darwinism, Bronn could still save the phenomena of progress that he spent so many years documenting and analyzing, by correlating morphological perfection with survival value.

This interpretation was not merely self-serving on Bronn's part. Darwin himself made liberal use of the word "perfected" in the original text, often interchangeably with "improved" or "favoured," so as almost to invite Bronn to link all three under a single term like *vervollkommnet*. To take just one example of many, Darwin described in a section entitled "Circumstances favourable to Natural Selection" the possible effects of repeated Lyellian geological cycles of continental subsidence and uplift. With subsidence, the continents became flooded and species were divided up into separate island populations, which might diverge from each other morphologically as they adapted to local conditions, forming geographic varieties. Then with uplift, the waters receded and the varieties, now again connected by dry land, would spread out and compete with one another. Inferior varieties might go extinct as a result, or else the competition might drive surviving varieties to diverge further from each other in character and become distinct species.

In Darwin's words, and with Bronn's translations of key terms in brackets, the process went as follows. After subsidence of the continent and creation of island populations,

New places in the polity of each island will have to be filled up by modifications of the old inhabitants; and time will be allowed for the varieties in each to become well modified and perfected [*um die Varietäten . . . umzugestalten und zu vervollkommnen*]. When, by renewed elevation, the islands shall be re-converted into a continental area, there will again be severe competition: the most favoured [*begünstigten*] or improved [*verbessert*] varieties will be enabled to spread: there will be much extinction of the less improved [*minder vollkommene*] forms . . . ; and again there will be a fair field for natural selection to improve still further [*zur ferner Verbesserung*] the inhabitants and thus produce new species.[39]

In English, the island populations were said to be more or less modified, perfected, favored, and improved. In German, Bronn used a similar mix of terms, but with preference for *Vervollkommnung* (for "perfection" but sometimes also for "improvement").

It was usually quite clear from context that the varieties that Darwin called "more perfect" were the ones that had become superior competitors, but why did he use such language at all? It seems to me that he was trying once again to co-opt the morphologists' standard terms for his own purposes. For example, as discussed above, Darwin had not rejected the morphologists' talk of higher and lower forms altogether, but had redefined the words in his own way to refer to relative competitive ability. In a similar manner, he appropriated and redefined "perfection" as well, making it refer to competitive improvement. But if Darwin could use "perfection" to mean what he wanted it to mean, without us labeling him a transcendentalist, why should Bronn not have used *Vervollkommnung*? Bronn's usage suited Darwin's purposes, too, because it appropriated the German morphological term for comparisons of competitive ability.

Bronn also wished to suggest that there was more to Darwinian progress and perfection than mere competitive improvement, that natural selection would also drive organisms up the various scales of morphological progress; and there were passages in the *Origin* that supported Bronn's reading. Darwin met Bronn more than halfway by playing up the importance of physiological division of labor—Bronn's most important morphological measure of progress and perfection—as a means of improving fitness. Darwin wrote, "No naturalist doubts the advantage of what has been called the 'physiological division of labour;' hence we may believe that it would be advantageous to a plant to produce stamens alone in one flower or on one whole plant, and pistils alone in another flower or on another plant."[40]

Darwin's second edition further confirmed Bronn in his interpretation, by adding another strong connection among selective advantage, physiological division of labor, and progress up the scale of nature: "The best definition [of morphological progress] is, that the higher forms have their organs more distinctly specialized for different functions; and as such division of physiological labour seems to be an advantage to each being, natural selection will constantly tend in so far to make the later and more modified forms higher than their early progenitors, or than the slightly modified descendants of such progenitors."[41]

And in the third edition, almost as if to endorse Bronn's reading of the second, Darwin changed the sentence to equate this form of evolutionary specialization explicitly to perfection: "We have also seen that as the specialization of parts and organs is an advantage to each being, so natural selection will constantly tend thus to render the organisation of each being more specialized and perfect, and in this sense higher."[42]

Division of labor was by no means the only way to gain competitive advantage, but it did provide the area of overlap between the Bronnian conception of progress and Darwinian fitness. Darwin made this very connection, and he strengthened it in successive editions of the *Origin*, perhaps for the purpose of drawing an older generation of morphologists into the Darwinian fold.[43] Even though the word has come to have some very un-Darwinian connotations, Bronn's usage of *Vervollkommnung* was not very different from Darwin's own usages of "perfection" and "progress," and was probably not unwelcome to Darwin. Bronn made Darwin speak the morphologists' language more consistently and say to them that scales of progress were to be defined and explained in new ways, but not rejected.

The Problem of Creation

One area where Bronn could not find common ground with Darwin was in connection with the mystery of mysteries: where new species came from. Bronn could not even begin to translate the title without choosing between two alternatives for "origin" that corresponded to their two contrasting conceptions. As Junker has also discussed, Carus later made an issue of Bronn's choice of *Entstehung* instead of *Ursprung*.[44] Both words could have been translated as "beginning" or "origin," and dictionaries often included the one in the definition of the other, but there was a difference in nuance that might have reflected one of Bronn's central reservations about Darwin's theory. *Ursprung*, related etymologically to the English "spring" or "wellspring," had connotations of something bursting forth, like water from a source,[45] and would have suggested an origin *de novo* more strongly. *Entstehung*, on the other hand, would have evoked a process of differentiating, developing, or arising out of preexisting components.

Why Carus disapproved of Bronn's choice is not at all clear, because one could argue just as easily that *Entstehung* better suited Darwin's purposes. Under Darwin's theory, species do indeed originate by gradual transformation of preexisting organisms, in contrast to successional theories like Bronn's that have species spring into existence. Perhaps Carus detected an implicit criticism in Bronn's title, for Bronn's wording, as Junker has suggested, called attention to Darwin's failure to deal with ultimate origins—the *Ursprung* of the very first species—to Bronn's satisfaction.[46] But then, again, since Darwin really did not try to explain ultimate origins, why should he have wanted *Ursprung* in the title? Carus was off the mark with some of his other criticisms of Bronn's German word choices as well,[47] so I suspect that he was just being overzealous in finding fault with his predecessor and currying favor with Darwin. In any case, Darwin expressed no opinion on the matter, other than to instruct Carus to retain Bronn's title for the sake of continuity with the earlier editions.

In his "Recapitulation and Conclusion," Darwin complained of unnamed "eminent naturalists" who clung to the idea that species were individually created, even while conceding some creative power to variation. Darwin criticized them with what for him was unusual sharpness, calling them "a curious illustration of the blindness of preconceived opinion," and in good, Paleyan style, he challenged them repeatedly to explain the phenomena as well as Darwin could himself:

These authors seem no more startled at a miraculous act of creation than at an ordinary birth. But do they really believe that at innumerable periods in the earth's history certain elemental atoms have been commanded suddenly to flash into living tissues? Do they believe that at each supposed act of creation one individual or many were produced? Were all the infinitely numerous kinds of animals and plants created as eggs or seed, or as full grown? and in the case of mammals, were they created bearing the false marks of nourishment from the mother's womb? Although naturalists very properly demand a full explanation of every difficulty from those who believe in the mutability of species, on their own side they ignore the whole subject of the first appearance of species in what they consider reverent silence.[48]

Bronn could not let such a challenge go unanswered, and he had every reason to consider it an unfair one. Bronn had never "ignored the whole subject," but rather tried to provide answers to at least some of these very questions. Here Darwin was the one ignoring the relevant parts of *Geschichte der Natur*. A footnote in Bronn's first edition of the *Origin* referred the reader ahead to the counterattack in Bronn's critical chapter, where Bronn threw the same questions back at Darwin. How did *he* picture the creation of the first few species? Was he not just as evasive as any of his opponents? As before, in his book review, Bronn argued that far from supplanting special creative forces, Darwin still needed some such agency himself, to "breathe life" into the very first organism or organisms. Darwin's formulation of this creative step varied from edition to edition, but the problem remained.

In his own first edition, Darwin had begun the history of life with "some one primordial form, into which life was first breathed." In the second, to which Bronn was responding, he specified "first breathed by the Creator," and Bronn took this additional nod to theism literally. Perhaps it was hasty of him to do so, since Darwin removed the metaphorical language from the third edition,[49] as Bronn later acknowledged,[50] but that was not enough to get Darwin off the hook or to make Bronn revise his critique. Bronn still saw in Darwin's account an implicit need, if not for divine Creation, then for some creative force no less mysterious than Bronn's.

Further, Bronn argued in his critical chapter, limiting the work of the creative force to just one or a few primeval specimens was not as easy or parsimonious as Darwin assumed. Darwin overlooked the principle of adaptation and the need for the first organisms to survive in their environment, as well as for the primeval world to function as a system. Again returning fire against Darwin's challenge to the successionists, Bronn posed similar rhetorical questions, with a similarly aggressive tone. He demanded to know "whether the first lichen, the first fern, the first palm and the first violet had life blown into them at the same time as the first infusorian, the first sea urchin, the first caterpillar, and the first frog, or whether it was done sequentially, together on the same spot, or on as many spots as there were species, spread out over the whole surface of the earth; and whether they began right away—insofar as they were within reach of one another—in the absence of other nourishment, to eat each other up, or what they lived on until they could reproduce?"[51]

Such ecological considerations convinced Bronn that a complex system of interdependent organisms would have had to originate as a unit. Once that point was granted, and the creative force was allowed to produce a number of different forms, then why should it not have been a large number, and why should it have acted only once instead of uniformly throughout the history of the Earth? The only alternative he could see for Darwin would have been a single simple alga as the first organism and a long period of time before it had filled the world with enough organic matter to support the simplest animals.

Either way, the creative force could not be gotten rid of, and it was all the same to Bronn, "whether the first creative act occupied itself with one or with 10 or with 100,000 species, and whether it did so once and for all or repeatedly from time to time. It is not a question of how many kinds of organisms it called to life, but whether it can altogether ever be necessary for it to interfere in the wonderful gears of nature and to help out in place of a motive law of nature?"[52] Even if Darwin had been able plausibly to limit its action to the production of a single algal cell, that would still not have rendered the creative force redundant or laid successional theories like Bronn's to rest.

This vigorous line of counter-questioning got Darwin's attention and elicited responses by letter and in later editions of the *Origin*. In one letter to Bronn, Darwin conceded that his account of the appearance of the first organisms suffered from the same shortcomings as the repeated creations of the successionists: "You argue most justly against my question, whether the many species were created as eggs or as mature &c; I certainly had no right to ask that question."[53] Accordingly, Darwin toned down his aggressive line of questioning in the next edition of the *Origin* (his third), where instead of accusing opponents of ignoring the problem of first appearances, he admitted that: "Undoubtedly these same questions cannot be answered by those who, under the present state of science, believe in the creation of a few aboriginal forms, or of some one form of life."[54]

Nevertheless, Darwin insisted to Bronn that even if he could not eliminate unexplained creative forces, there was virtue in circumscribing their actions. Darwin's combination of very few original species with descent of all the rest from them provided a superior explanation of why plants and animals could be classified into a small number of major types: "I fully agree that there might have been as well 100,000 creations as 8 or 10, or only one. But then, on the view of eight or ten creations (i.e. as many as there are distinct types of structure) we can on my view understand the homological & embryological resemblance of all the organisms of each type; & on this ground almost alone I disbelieve in the innumerable acts of creation."[55]

Along the same lines, he added, in his third edition, that it was also more parsimonious to assume that nature used the creative forces only as much as necessary, and that she would not have created misleading evidence of common ancestry: "It has been asserted by several authors that it is as easy to believe in the creation of a hundred million beings as of one; but Maupertuis's philosophical axiom 'of least action' leads the mind more willing to admit the smaller number; and certainly we ought not to believe that innumerable beings within each great class have been created with plain, but deceptive, marks of descent from a single parent."[56]

A New Newton of Biology?

One more argument of Darwin's must certainly have struck Bronn as misdirected. Bronn had devoted most of his career to the scientific illumination of his paleontological data, in search of patterns and laws of succession, undeterred by his inability to explain the origin of new species. Yet Darwin felt that he himself was the better Newtonian for describing the course of evolutionary history, rather than feigning hypotheses about what life was or how it originated. He rehearsed the argument in a letter to Lyell:

With respect to Bronn's objection that it cannot be shown how life arises, & likewise to a certain extent Asa Gray's remark that natural selection is not a vera causa,—I was much interested by finding accidentally in Brewster's 'Life of Newton,' that Leibniz objected to the law of gravity because Newton could not show what gravity itself is. As it has chanced, I have used in letters this very same argument, little knowing that any one had really thus objected to the law of gravity.—Newton answers by saying that it is philosophy to make out the movements of a clock, though you do not know why the weight descends to the ground.—Leibniz further objected that the law of gravity was opposed to Natural Religion!—Is this not curious? I really think I shall use the facts for some introductory remarks for my bigger book.[57]

To Bronn, Darwin wrote: "Lastly, permit me to add that I cannot see the force of your objection, that nothing is effected until the origin of life is explained: surely it is worth while to attempt to follow out the action of Electricity, though we know not what electricity is."[58]

Bronn answered that Darwin's analogy was far from apt. He said that electricity, like gravity, was an inherent property of certain forms of matter, and the physicist was not expected to give an account of how the property originated or how matter acquired it. Matter and its properties were inseparable and the physicist studied them together. Life, in contrast, did not inhere in all organic matter in the same way. It was indeed something separable from organic matter, and it therefore required a separate explanation. We could observe life or life force being transferred (within a species) during reproduction, and so we needed to ask how the species first acquired it.[59]

Darwin missed Bronn's point. If he had thought about Bronn's approach, Darwin would have seen that Bronn, too, had put aside the question of *Ursprung* in order to "follow out" the course of organic history and formulate laws of survival and progress and did not need a lesson on Newtonian methodology. Bronn's point was not that "*nothing* is effected until the origin of life is explained"; it was that Bronn's theory had not yet been superseded, because Darwin had not completely replaced creative forces.

Bronn saw Darwin's theory as a hybrid, combining the special creation or production of some arbitrary number of early species, combined with evolutionary origins for all the others, and he saw both advantages and disadvantages to the evolutionary part of the story. The big disadvantage was that gradual Darwinian evolution should have produced a chaos of unclassifiable transitional forms, much worse than what taxonomists actually had to deal with. The burden was on Darwin to explain why the systematist in fact observed such a chaos neither among good species at any one time nor among successive fossil species. Darwin's explanation was, of course, that natural selection tended to carve distinct varieties and species out of the myriad variations. Over geological time, transitional forms were replaced so rapidly by better-adapted descendants that they rarely showed up in the fossil record, which was incomplete and unreliable anyway. But this solution was, again, not good enough for Bronn.

Bronn rejected the arguments about the incompleteness of the fossil record, which was ironic, because Darwin's chapter on the subject was largely in agreement with Bronn's own analysis in *Geschichte der Natur*. But Darwin's chapter cited Bronn only once, and then only to question the relevance of one of Bronn's less-important observations to the effect that the average sedimentary rock formation lasts two to three times as long as the average species.[60] Bronn must have been irked by the offhand treatment, because he retaliated in his concluding chapter by taking Darwin to task for his inadequate analysis of the fossil record.

At issue, for Darwin, was whether one should have expected the fossil record to be fairly complete for any segment of time and space captured within a single, continuously deposited formation, and how often species' life spans lay entirely within one such formation. Darwin thought that if a species appeared abruptly in the middle of a formation and disappeared just as abruptly before the top, then, even if it was abundantly preserved in between, one still would not be justified in assuming that its whole history had been recorded. Transitional forms still could have been missed. Bronn countered in his critical chapter that so many different species were found in such great numbers in so many identical formations that it was hard to believe that transitional forms would *never* be among them. Darwin, Bronn acknowledged, could supply ad hoc explanations for any number of individual cases of missing transitional forms, but he could not possibly account for the consistent, lawlike absence of them everywhere: "Conditions for the preservation of transitional forms could not possibly have been so completely unfavorable that absolutely nothing of them remains."[61]

To illustrate the problem of order in the present, Bronn developed a counterexample of two species of rat, which I will explain fully, because Darwin later responded to it. Bronn referred to the two species as the house rat, *Mus rattus* (=*Rattus rattus*, also known as the black rat or ship rat) and the sewer rat, *Mus decumanus* (=*Rattus norvegicus*, also known as the Norway rat, brown rat, water rat, or laboratory rat). The sewer rat was observed to be displacing the house rat in Europe, but why? And how did Europe end up with precisely these two as the final competitors, among all the rat varieties that might ever have existed? The two rat species differed in at least four easily recognized characteristics: size, coloration, ear-length, and tail-length. Hence, as Bronn calculated very roughly, ancestral rats could have had any of tens of possible combinations of each of these traits with the others, and that was without even counting the large number of intermediate characteristics that might or might not have occurred in the past.

Given so many possible ways of putting together a rat, Bronn demanded to know why precisely the currently observed sewer-rat combination emerged as the superior competitor. Why should he believe, in addition, that every step of the evolutionary transition from the ancestral rat to the sewer rat conferred a competitive improvement? And why did this superior sewer rat first have to beat out all of its partially improved intraspecific rivals and become a well-defined species, before it started displacing the house rat? Surely some of the intermediate forms should have lasted longer in the struggle than the inferior house rat. They should still have been around, if they had ever in fact existed.

Bronn had been studying Darwin closely enough to anticipate the answer. Darwin, Bronn predicted, would have cited the many uncertainties surrounding the appearance and distribution of the ancestral rat species, the sequence of intermediate forms, and the precise utility of the various combinations of traits. He would have said that his argument did not depend on the precise course of the story. All that really mattered was that some form of sewer rat emerged that had eliminated all of its immediate competitors and was now encroaching on the house rat. In short, Darwin would have weaseled out of his obligation to provide historical details, as he did in most other cases as well. For example, among blind cave-dwelling animals, Bronn asked, why did intermediate, half-blind varieties not survive in the twilight near the front of the cave? Darwin said they must have been outcompeted. But did that necessarily happen? No matter what we observed, Darwin could tell one story or another to explain it in terms of natural selection.[62]

Bronn's problem with such hypothetical historical narratives was not that they were implausible or even unlikely, but that they were uncertain:

In response to every objection, he [i.e., Darwin] will hold out this and similar general answers, which taken for themselves are unassailable; but even if they are well founded in some *individual* cases, and can in no case be dismissed as absolutely unfitting or refuted, still everyone must feel that the matter on the *whole*, even following the Darwinian theory, would have had to turn out and would still have to turn out quite differently from what is in reality the case. He has the advantage that he does not owe us an account of *any one* individual case, because we cannot reasonably demand an account of *every* individual case from him![63]

Bronn was willing to give some credence to Darwin's hypothetical histories and to concede that natural selection could have and probably would have maintained order and species' identities in these ways in any number of individual cases, but he still saw no logical or natural necessity for it always to do so: "But may this explanatory device really be considered the rule, and its inapplicability only the rare exception?? Must it in all cases have been so, just because it could have been so in individual cases?"[64] Here again, we see the conflict of Bronn's conception of *Wissenschaft* and ideal of formulating general and necessary laws with the characteristically Darwinian strategy of providing hypothetical historical narratives to explain individual cases.

In response, Darwin acknowledged the uncertainty involved in reconstructing individual cases, but not the need to seek general laws of organic history. He continued to challenge his opponent to provide a better explanation than the evolutionary one: "You put very well & very fairly that I can in no one instance explain the course of modification in any particular instance. I could make some sort of answer to your case of the two Rats; & might I not turn round, & ask him, who believes in the separate creation of each species, why one Rat has a longer tail or shorter ears than another? I presume that most people would say that these characters were of some use or stood in some connection with other parts; & if so, natural selection could act on them. But as you put the case, it tells well against me."[65]

Here Darwin seemed once again to be arguing past Bronn and taking on imagined British rivals instead. Unlike the natural theologians, Bronn was not fully committed to the utilitarian design of species or the idea that new characteristics had to be "of some use," or correlated with something of use. In Bronn's system, a new rat species might have gained a longer or shorter tail in fulfillment of some other law of morphological progress.

Still, Bronn's point was not entirely lost on Darwin. There are several places in his correspondence where he discussed the problems of how to reconstruct individual case histories and how to judge whether a characteristic was an adaptation, and the subject came up even after Bronn's death. In 1863 he still seemed resigned to being unable to answer Bronn definitively: "When we descend to details, we can prove that no one species has changed: nor can we prove that the supposed changes are beneficial which is the groundwork of the theory. Nor can we explain why some species have changed & others have not. The latter case seems to me hardly more difficult to understand precisely and in detail than the former case of supposed change. Bronn may ask in vain the old creationist school & the new school why one mouse has longer ears than another mouse—& one plant more pointed leaves than another plant."[66]

With time, however, Darwin came to seem more assured of the adaptive value of even trivial characteristics. He engaged a correspondent in 1875 in a detailed discussion of mouse tails and the important functions they are likely to serve, contrary to Bronn's assumptions.[67] And by 1880 he expressed satisfaction to Huxley that the preponderance of the evidence had shifted to the adaptationist side:

If I think continuously on some half-dozen structures of which we can at present see no use, I can persuade myself that Natural Selection is of quite subordinate importance. On the other hand, when I reflect on the innumerable structures, especially in plants, which twenty years ago would have been called simply "morphological" and useless, and which are now known to be highly important, I can persuade myself that every structure may have been developed through Natural Selection. It is really curious how many out of a list of structures which Bronn enumerated, as not possibly due to Natural Selection because of no functional importance, can now be shown to be highly important.[68]

One last interpretive issue that continually played a prominent role in the reception of Darwin's theory was the extent to which it substituted chance for developmental law or divine design, thus depriving the Creation of its harmony and purpose. Other critics, in Germany and elsewhere, considered this one of the most objectionable features of Darwinism, but Bronn took it in stride. He ascribed very little significance to the matter, listing it among the many lesser problems that had to await further research and future judgment.

Bronn spent more time introducing the problem than discussing it, as if to reassure the reader that he did not lack in appreciation of nature's wonders. He recited a litany of ways in which a morphologist could discover plan and purpose in nature:

The more a naturalist has occupied himself with the minute study of the structure of natural beings and their wonderful purposefulness, of the coordination of all the details in an organism, where no particle can be changed indiscriminately without endangering the whole,—of the repetition of the same planned features, always in unique ways, in the 250,000 known species of the present period of creation,—of the culminating consummation of the whole in the most perfect of these organisms,—of the development of all these features for future purposes in an embryo that does not yet need them—the harder it will be for the naturalist, at first, to see nothing further in all this than the consequences of a progressive process of improvement, in which every additional new step forward, according to the author's [i.e., Darwin's] theory, is always just a *chance occurrence* [*ein Zufall*] and can only be preserved by heredity.[69] (Emphasis in original)

But then Bronn suddenly shifted his perspective, and simply dismissed any concern about Darwin's demystification of life: "And, yet, one must not see in this any unconditional obstacle to the theory!"[70] If Darwin was right about everything else, traditional conceptions of harmony, plan, and purpose would just have to be abandoned.

Conclusion

Bronn's understanding of Darwin's work was inevitably colored not only by linguistic ambiguities, but by contrasts between the two authors' stores of knowledge and experiences in natural history, pure and applied; their contrasting social roles as professional researcher and self-supporting gentleman; differences between German and British conceptions of biology as a science; and from Bronn's side, at least, by rivalry between an established theorist and a newcomer to the field.

Bronn consistently applied the methodological ideals of university-based German *Wissenschaft* to the history of nature. He stressed quantification of change and progress, and coordination of data from around the world and across geological periods. He worked to discover deterministic, necessary laws governing an orderly organic world. He strove to unify the life sciences with physics, geology, and ultimately cosmology, by connecting organic change to the physical evolution of the Earth, and by placing his laws of life in a hierarchy of physical and chemical laws. Darwin, the gentleman-naturalist, trained by natural theologians and skeptical of the all-sufficiency of blind laws or progress, did not work by the same rules as Bronn, or try to appeal to the same academic audience. He did not feel the need to assert and uphold the scientific status of biology or to embed his theory of organic change in a larger system of laws and an account of ultimate origins. Darwin was particularly un-*wissenschaftlich*, by Bronn's standards, in his manner of piling up individual examples and devising explanatory narratives of what might have happened in the highly contingent evolutionary past, instead of deriving necessary laws and sequences of historical events.

Nonetheless, Bronn was prepared to accept the revolutionary biological, philosophical, and theological implications of Darwin's theory, if only Darwin's argument could ever be shored up sufficiently to pass muster as a *wissenschaftlich* one. In the meantime, Bronn thought it was important to inform German readers about Darwin's proposals and provide them with not only a translation, but also a guide to the issues that Darwinians would need to address in the coming years. He also tried to lead German readers toward an understanding of Darwin's theory that made it as compatible as possible with his own system of organic history.

Parts of Darwin's argument were weakened or challenged in the translation. The significance of Darwin's extended analogy to artificial selection was not fully conveyed. Bronn had little interest in or appreciation for Darwin's use of the analogy to establish selection as a Herschelian *vera causa* that could take the place of a law or force of nature. Darwin's strategy of coaxing the reader gradually toward his view of nature by starting the book with familiar examples from British farms, kennels, and pigeon coops, was wasted on Bronn, who was neither familiar with the breeds nor convinced of the analogy between artificial improvement and natural progress and perfection. And Darwin's metaphorical move from the artificial breeder to a personified natural selection failed similarly. The rhetorical connection to Paley's argument from the artificial to the natural made no discernible impression. Bronn would much rather have seen natural selection as a proper, deterministic law of nature than an unpredictable agent.

Still, Bronn discovered that he had much in common with Darwin. They both cast their nets broadly to encompass organic change on all scales, from the global and biogeographical to variation in the field and the garden. They both focused on the problem of how species could always be well adapted in a changing physical and biotic environment, and both looked for the causes of faunal and floral change in the environment. But Bronn was always aware of the incompleteness of his system, because although he could easily explain how the changing environment drove existing species to extinction, he could not fully account for the origin of better-adapted ones. He could only describe the patterns and laws of change, and he was tantalized by Darwin's mechanism of natural selection, which made adaptive change plausible and necessary, even if it was not supported by a complete and well-structured *wissenschaftlich* argument.

Bronn's translation, including the chapter of commentary, was better than its reputation. Darwin's account of variation in nature, Malthusian increase and struggle for existence, and differential survival and reproductive success came through clearly. Bronn was also very clear and approving on the role of adaptation to a continually changing environment as the driving force behind historical progress. He found support in it for his own ideas about the logical and natural necessity of adaptation, and it removed the ultimate cause of change from the biological realm to the external environment, where Bronn himself had always located it.

Bronn also emphasized Darwin's consistent naturalism and elimination of design and purpose from biology (though perhaps not from nature as a whole), as well as the consilience of Darwin's inductions and his remarkable ability to solve disparate problems in biogeography, paleontology, taxonomy, embryology, and other areas. He was especially interested in Darwin's approach to defining and explaining the historical increase in morphological complexity and "perfection" in terms of physiological specialization and adaptation.

Bronn took it upon himself to make Darwin's theory better known in Germany, but not because he saw it as derivative from or compatible with older German ideas to which he still clung. For a critical, yet constructive introduction of his ideas into Germany, Darwin was indebted not to the old *Naturphilosophen* or conservative, idealistic morphologists, but to the innovative Bronn, who had already rejected the constraints of transcendental archetypes, recapitulation, and developmental analogies, and who was looking for new solutions to problems he himself had been grappling with for as long as Darwin.

Throughout his two years of contact with Darwin, Bronn remained ambivalent about Darwin's theory. "I have probably read a dozen critiques of your work in German, Dutch, English, and American journals, favorable and many unfavorable," he wrote in 1862, in what are probably his last recorded words on the subject. "All this while, they have changed nothing in *my* judgment"[71] (emphasis in original). He now singled out the problem of ultimate origins as the main one. He saw in Darwin's theory "the only naturalistic way to finally solve the mystery of Creation," but could find insufficient empirical support for Darwin's account of the first species.

To anyone wishing to establish Darwinian evolution in Germany, Bronn bequeathed a solid foundation on which to build, but also a number of explicit challenges to overcome. Bronn's desired equivalence between fitness and morphological progress had to be better established. Bronn's species concept, with its limits on the accumulation of variation, had to be revised. At the same time, however, variation had to be reined in so as not to create a chaos of transitional forms. Most of all, Darwin's account of life being breathed into the very first species had to be made more precise, naturalistic, and plausible, and his overreliance on hypothetical narratives in the absence of hard historical evidence and necessary laws of change had to be remedied. As we shall see, these are the challenges to which Ernst Haeckel devoted his life's work.

5

Ernst Haeckel as a Darwinian Reformer

Into the vacuum in German Darwin interpretation, left by Bronn's death in 1862, stepped the young Ernst Haeckel, who did more than anyone outside England to promote, defend, and elaborate upon Darwin's theory, to unify biology within a historical framework, and to design a practical program of research into phylogeny. This chapter explores Haeckel's debts to both Darwin and Bronn, his answers to Bronn's challenges, and his differences from the older morphological tradition, which consistently have been underestimated in the historical literature. Like Bronn, Haeckel might have used the language and metaphors of development, spoken of laws and forces of *Entwicklung*, and juxtaposed the development of the embryo with the history of the species, but the words no longer had the same meanings as before 1860. Haeckel's new uses of *"Entwicklung"* referred to an open-ended and unpredictable process, not a chain of fully predetermined forms. When he equated traditional types with Darwinian ancestors or lineages, he was not misunderstanding or undermining the Darwinian conception, but indicating to morphologists and systematists where and how to apply it. Phenomena such as unity of type were explained in new ways, even if they were not given new names. Similarly, when Haeckel spoke of progress and perfection, he referred to historical trends like those Bronn had described and analyzed, not to an idealized sequence of forms, and he explained them in terms of adaptation and selection.

Haeckel's laws, like Bronn's, were usually empirical generalizations about historical patterns and tendencies. He did not present them as the causes of those patterns, he did not deduce the existence of vital forces from them, and he did not believe that they compelled evolution or development to follow predetermined linear pathways. Rather, they interacted in complex ways to produce variation, diversity, and novelty, in the same manner as Bronn's laws. Haeckel formulated so many laws of evolution, in fact, each with subsidiary laws, counterlaws, law-governed exceptions, and other escape clauses, that the results were very un-law-like and unpredictable—as unpredictable and historically contingent, I would add, as Darwin intended variation and evolution to be. All of Haeckel's talk of law should not obscure the fact that Haeckel's system expressly ruled out teleology, divine providence, and any special biological determinism. Concerning Haeckel's "Lamarckism," I argue that it, too, was more rhetoric than substance, for he declared that most of it had been superseded by Darwin, and he rejected basic features of contemporary Lamarckism, such as the linear animal scale and animals' inner drive to ascend it, or the active role of the animal psyche in responding to perceived needs and thereby shaping the body. The Lamarckian drive to perfection was far too similar to German transcendentalist assumptions to be acceptable to him, and the psychological factors, by at least some interpretations, were too "dualistic," in the sense of allowing for a mental realm that was independent of the physical body and brain. Haeckel's was a Lamarckism stripped of the above features, but retaining inheritance of acquired characteristics and adding universal common descent. Even the inheritance of acquired characteristics was decidedly un-Lamarckian in Haeckel's hands, because it did not produce fully formed adaptations all by itself. The Lamarckian mechanism produced heritable, mostly favorable variation, upon which Darwinian natural selection still had to act. Darwin had used what we now think of as "Lamarckian heredity" (use-inheritance and environmental effects) in the same manner, even though he might have put greater emphasis than Haeckel on "individual differences."

After reading Bronn's *Origin* in 1861, it took Haeckel another five years to work out the complete system of evolution, morphology, systematics, and monist philosophy that he detailed in his magnum opus, *Generelle Morphologie der Organismen* (General morphology of organisms). During that time, he had to shore up Darwin's theory against Bronn's criticisms and against any suggestion of an initial Creation or predetermination of the course of evolutionary history. He also had to launch his own career as a zoologist and establish his credentials, first as a systematist, then as a follower of Darwin, and finally as a Darwinian theorist in his own right.

Haeckel Discovers Darwin

By the turn of the twentieth century, Haeckel's many admirers could look back with pride at the moment in September 1863 when he stepped to the podium to address the German Society of Naturalists and Physicians (*Gesellschaft deutscher Naturforscher und Ärzte*—hereafter *GDNÄ*) and usher in the era of Darwinian biology in Germany. He was the first speaker at the meeting in Stettin (since 1945, Szczecin, Poland) and his performance was stellar. Haeckel's early biographer and popularizer, the nature-writer Wilhelm Boelsche, emphasized the effects of his youthful magnetism on the audience: "One must remember what magic emanated from his person, speaking just of appearances. It was the kind of immediate magic that did not need to take an indirect route through his nascent zoological reputation."[1]

Boelsche thought Darwinism was fortunate to have found in Haeckel a much better-looking figurehead than Darwin himself, who could never have created such a sensation:

Darwin was never a handsome man in the ideal sense. . . . In all the years he was writing the "Origin of Species," he did not even wear the patriarch's beard that now seems to us inseparable from his countenance: the crown was already bald, but the chin shaven smooth. The prematurely bent figure of this dyspeptic man could never, for all its venerability, have had the same effect in that situation. With Haeckel's youthful beauty came something like the incarnation of the old ideal of the "*mens sana in corpore sano.*" . . . With him came the best thing that can accompany a new idea: the air of a new generation, a youthfulness that brings with it the rosy courage for new ideas altogether.[2]

At twenty-nine, Haeckel was indeed relatively young to be opening the first plenary session with a controversial talk, and his youth was not an unambiguous asset. His opponents called attention to his youthful enthusiasm as evidence of his inexperience, especially in geology and paleontology.[3] Still, Haeckel's star clearly was rising, and he was perceived as a young pioneer of a fundamentally new approach to the life sciences.

Haeckel had been born in 1834, in Potsdam, Brandenburg, in what was then the Kingdom of Prussia. The son of a state-government official, Haeckel had followed his father's wishes and studied medicine, even though he would much rather have become a botanist. Nevertheless, he found ways to pursue his interests in natural history and biological theory during his medical studies at Würzburg and Berlin. He learned cell biology from Rudolf Virchow, marine invertebrate zoology from Johannes Müller, and histology and anatomy from Albert von Kölliker. He also made the acquaintance of his future collaborator Carl Gegenbaur, who studied with Müller in Berlin and completed his *Habilitation* in Würzburg when Haeckel was there.

By 1858, Haeckel's dissertation, medical studies, and licensing exams were all finished, and although he was determined to pursue a career in scientific research, he dutifully hung out his shingle in Berlin first, to please his father. However, he did not feel obligated to go out of his way to attract patients. He later joked, "I can certify at least one thing about this medical practice of mine: that not one of my patients died. I also held my office hours from 5 to 6 o'clock in the morning, so I only had three patients the whole time." This count apparently included his parents' chambermaid, whom they sent to him once to have a finger examined. His parents got the message: "This much success was enough for my dear father, and in 1859 he let me go to the land of my heart's desire for a year, to Italy."[4]

The transition back into zoological research had not been quite as easy as Haeckel made it sound, however. Haeckel originally planned to continue his studies toward a *Habilitation* under Johannes Müller, whom Haeckel idolized, but when Müller died suddenly in 1858, Haeckel was disconsolate and uncertain how to proceed with his career. He got timely encouragement and direction from two sources that year, his cousin Anna Sethe to whom he became engaged, and Gegenbaur, who began to recruit his promising fellow Müller-acolyte for an expected job opening at the University of Jena. As part of his campaign, Gegenbaur asked Haeckel to accompany him on a planned research trip to Italy. Haeckel was overjoyed at this opportunity to restart his scientific career and went ahead with the trip in January 1858, even though Gegenbaur had to back out because of illness.[5]

In Italy, Haeckel decided to follow up some of Müller's late work on the Radiolaria, a group of single-celled organisms with intricate, glassy skeletons. He collected and described many new genera and species and developed a natural system of classification for them as a unified group, based on their skeletal morphology. The result was a monograph, dedicated to Johannes Müller and beautifully illustrated with copper engravings. It also contained his first published argument in favor of Darwinism. He had read the *Origin*, in Bronn's translation, after his return from Italy, and he kept it in mind as he wrote up his Radiolaria results.

The brief pro-Darwinian argument was uncharacteristically tentative compared to Haeckel's later writings, and most of it was relegated to a long footnote, but it declared (twice) the beginning of a "new epoch" in the study of organic nature. Haeckel already foresaw that the investigation of evolutionary relationships and transitional forms between taxonomic groups could lead the way toward unifying the biological sciences and ridding them of teleology and religion. He praised Darwin's work for being "the first serious, scientific attempt to explain all the phenomena of organic nature from a grand, unified point of view and to replace incomprehensible wonder with comprehensible law of nature. . . . What is being made here is, of course, only the first great attempt to even aim towards a scientific, physiological history of creation of the organic world at all and to prove that the physiological laws and the chemical and physical powers that the world of present-day Creation obeys also ruled in the prehistoric world."[6]

Haeckel also echoed some of Bronn's ambivalence about Darwin's theory, but not on the questions of species transformation and common descent per se, which he clearly accepted right away. He only hinted at some problems of evidence and at the possibility that struggle, selection, and all the rest of Darwin's principles might not exhaust the causes of organic change. On the positive side, Haeckel ventured to answer some of Bronn's objections, particularly the claim that Darwin had failed to eliminate the need for a special Creation or formative force. Even if Darwin did leave open the possibility that the very first species was created, Haeckel said that such an inconsistency in the otherwise perfectly naturalistic account should probably not be taken literally. Indeed, even at this early stage, Haeckel made it clear that the greatest attractions of Darwin's theory for him were its freedom from teleology and divine intervention, and its promise of substituting scientific understanding for superstitious wonderment. Despite all the difficulties that remained to be overcome, Haeckel endorsed the positive half of Bronn's two-minded conclusion. He quoted it at length, to the effect that Darwin was at the very least on the right track toward a naturalistic understanding of morphology, development, and evolution: "With the translator Bronn I see in Darwin's direction the *only possible way* to come close to understanding the great law of development that controls the whole organic world"[7] (emphasis in original).

Haeckel did not insert Darwin's theory gratuitously into his monograph simply because it was new and exciting. Haeckel had an immediate use for it in interpreting his observations on the Radiolaria, and he saw an opportunity to lend empirical support to Darwin's case. The Radiolaria, being single cells or simple colonies, did not have any appreciable embryological development of the sort Haeckel would later be concerned with. Here he was engaged in a project of comparison and classification of fully developed forms and showed no interest in Darwin's arguments about embryology or even types and adaptation. It was the chapter entitled

"Variation in Nature" and Darwin's argument about a species manufactory that first got Haeckel's attention and converted him to Darwin's
point of view. Haeckel's remarks on variation in the Radiolaria should
give pause to anyone who thinks he had a vested interest in preserving the
type concept or that he approached evolution strictly from the point of
view of embryology.

Haeckel introduced his first contribution to Darwinism, in the main
text, with some observations on the frequent occurrence of intermediate forms in his collections, and the gradual nature of the morphological transitions between groups: "I cannot leave this general depiction of
the relationships among the various Radiolarian families without having
especially emphasized the numerous *transitional forms*, which connect
the various natural groups most intimately, and which make their systematic division in part very difficult. If one considers what a small portion
of the entire Radiolarian fauna might so far have even been exposed to
our view at all, then it must seem doubly important and interesting that
already in this array of forms a fairly unbroken chain of related links can
be put together."

Haeckel saw in the continuous variation of his Radiolaria a possible
source of support for Darwin's theory:

I wish all the more to call attention to these intermediate links, because just
now every contribution must seem especially welcome that can provide special,
detailed research toward deciding the question of the gradual development of
organic beings from common ancestral forms. The great theories that *Charles
Darwin* recently developed "on the origin of species in the animal and plant kingdom by means of natural selection or the preservation of the perfected races in
the struggle for life" [my back-translation of Bronn's title; Haeckel is not going
by the original English], and with which a new epoch has begun for systematic
organic natural science, have suddenly given such significance to the question of
relationships among the organisms and such a fundamental importance to the
demonstration of a continuous linking, that even the smallest contribution that
can play a role in the further solution of these problems must be welcome.[8]

Ideally he would have liked to look for intraspecific variation, but had
compared too few individuals per species to make a good survey: "As
far as the question of the transformability of species itself is concerned,
my Radiolarian studies can make only a few contributions, which can of
course easily be explained by the relatively very small number of individuals that I was able to study and measure comparatively, because of the
short time and the overwhelming richness of the materials in Messina. In
the best case, I have investigated just a few 100 individuals of one and the
same species, and such a number is naturally much too low for such pur-

poses."⁹ His data did allow him, however, to test an important prediction of Darwin's about the distribution of varieties and species within genera: "Yet, I believe that I have noticed, especially in the Acanthometrans, multiple transitions, and therein found a verification of Darwin's claim that particularly *the largest genera tend the most toward modification of the species*"¹⁰ (emphasis added).

Haeckel then listed some of the species among the Acanthometrans and other families, among which he found intermediate forms. These groups, he argued, were rich in species and the species were highly variable, as Darwin predicted. Haeckel inferred from this that species were not fixed and that not only the species, but every taxonomic ranking, was changeable and subjective: "This last example [*Collosphaera huxleyi*] seems more than many other variety-rich species to be appropriate for providing a conception of the changeability of species, as well as of the shakiness of the species concept, which at a fundamental level is no less abandoned to arbitrary abstraction and subjective opinion than the concept of the individual, the concept of the genus, family, order, etc."¹¹

We see from the end of this quote that Haeckel was already looking for ways to undermine the reality and fixity of all the taxonomic categories, presumably including the highest classes and types. He welcomed Darwin's argument that, as his manufactory yielded new subgroups, the groups containing them rose in taxonomic rank to become species, genera, and so on, and that every level of the system was in flux.

Haeckel and Gegenbaur

The Radiolaria work earned Haeckel his *Habilitation* at Jena in 1861, where he immediately began to lecture on zoology. Within a year, Gegenbaur had engineered Haeckel's promotion to extra-ordinary professor and his installation as director of the zoological museum. With his living thus secured, he married Anna Sethe. Their future looked rosy, as Haeckel set his sights on a career as a Darwinian evolutionist. His earliest lectures on zoology at least mentioned Darwin, but from the 1862–1863 academic year on, Haeckel broke new ground with a course on the Darwinian theory, including a treatment of human evolution. Darwin had touched upon the subject only obliquely in the *Origin*, in a passage that had not even appeared in the German. The lectures therefore made original contributions to Darwinism and its applications, and they began to establish Haeckel as an authority, and as a very popular lecturer.¹² At the same time, Haeckel began his collaboration with Gegenbaur on a program of evolutionary morphology, aiming to establish historical relationships among all the major animal groups. The partnership lasted until 1873, when Gegenbaur followed a call to the University of Heidelberg.

Almost everyone who writes about the two Jena evolutionists follows William Coleman in ascribing greater seriousness and intellectual sophistication to Gegenbaur than to his younger partner, making Gegenbaur essentially the brains behind Haeckel.[13] But Gegenbaur's interests in evolution were also narrower than Haeckel's, and one misses a great deal of Haeckel's program by focusing just on his areas of overlap with Gegenbaur and dismissing everything else as flighty and fanciful. Further compounding the interpretive problems is the historical accident that Gegenbaur published his major work on morphology, the *Grundzüge der vergleichende Anatomie* (Elements of comparative anatomy), in 1859, before he read *On the Origin of Species*.[14] Gegenbaur's 1859 views, especially his concern at that time with type and homology, therefore hang like an albatross around his and Haeckel's necks, because Gegenbaur did not get around to revising and Darwinizing his book until 1870. Even then, he retained the original emphasis on types, rather than introducing new discussions of variation, selection, and adaptation. The conceptual framework of 1859 and the preoccupation with types is presumed to have undergirded and motivated their work of the 1860s, which is when Haeckel made his major statements on evolutionary theory, and to have established the character of German evolutionary morphology for the rest of the century, if not longer.

In neither edition of his book did Gegenbaur delve as deeply as Haeckel into questions about the causes of evolutionary change—the mechanisms of variation, heredity, and selection. The focus in the secondary literature on his treatments of types may therefore be quite appropriate. My concern here is not so much to raise new questions about Gegenbaur as to divorce Haeckel's work of the 1860s from Gegenbaur's of 1859. Haeckel was not beholden to Gegenbaur (or Cuvier or transcendentalism) for his idea of type. Moreover, Haeckel was the more complete Darwinian of the two collaborators, in the sense that he concerned himself not only with historical reconstructions and types, but with questions of evolutionary mechanism as well.

Coleman does note in passing that there was an interplay of heredity with adaptation in Haeckel's "simplified conception" of evolution, and that Gegenbaur allowed for such an interplay as well. Still, Coleman does not think that that side of their evolutionary morphology was very important to them, or to anyone else, because Gegenbaur's 1870 edition did not bother to develop it. Coleman is not alone in this assessment. Bowler's account of evolutionary morphology likewise assumes that reconstructing phylogeny and relationships was the main concern of morphologists through the late nineteenth century and their only work of lasting value. Their ideas about mechanisms of variation and change were of much less interest. Di Gregorio seems to concur.[15]

But matters of evolutionary theory and especially the mechanistic causes of variation and change were of the utmost importance to Haeckel. His monism and his arguments against teleology, religious dogma, and Creation relied heavily on the historical contingency and creativity of evolutionary change, and it made a great difference to him how variation was generated and preserved by selection and heredity. We have also seen from the Radiolaria monograph that Haeckel was initially attracted to Darwin's arguments about variation, and that he immediately embraced Darwin's destabilization of types and taxonomic ranks.

The interpretation of Haeckel as a typological thinker has become a self-confirming one. Any time Haeckel uses the word "type" he is read as a reactionary, and any time he uses "common ancestor," "phylum," or "heredity" he is presumed actually to mean "type." All of his interpretations of Darwin are spun into defenses of the type concept. To take just one egregious example, consider the following quote from a notebook on the *Origin* that Haeckel kept in the early 1860s, as transcribed and interpreted by Di Gregorio. It appears to Di Gregorio to fall short of full acceptance of the consequences of Darwin's theory:

216, 217. The two great fundamental laws of the *formation* of organic beings are:
I. Unity of type, that is, agreement in fundamental structural plan (*Naturphilosophen*). This is nothing other than unity of descent: heredity!!
II. *Adaptation to the conditions of existence* of organic and inorganic nature (Cuvier).—This explains itself through Natural Selection!! And this, by virtue of the inheritance of former adaptations, includes in itself the law of unity of type! Both laws together generate the diversity [*Mannichfaltigkeit*] of organic Forms. (Emphasis in original; my translation)

[216, 217. die beiden großen Grundgesetze der Bldg organischer Wesen sind:
I. Einheit des Typus also Übereinstimmung im Grundplan des Baues (Natur-Philosophen). Diese ist nichts anders als Einheit der Abstammung: Erblichkeit!!
II. *Anpassung an die Existenzbedingungen der organischen* und unorganischen Natur (Cuvier).—diese erklärt sich dch [=durch—M. di G.] Natürliche Züchtung!! Diese begreift vermöge der Erblichkeit früheren Anpassungen des [*sic*—S. G.] Gesetz der Einheit des Typus in sich! Beide Gesetze bilden zum [=*zusammen?*—S. G.] d Mannichfaltigkeit d. organischen Formen.][16]

Di Gregorio reads this as a confirmation of the thesis that from the very beginning of his career as an evolutionary morphologist, Haeckel was only reforming the type concept, not fully embracing Darwin's theory. He takes Haeckel's point (I) to show Haeckel connecting heredity and common descent back to the old type concept, thereby salvaging the latter. But this is not plausible, because the quote alludes to the *Naturphilosoph*'s types, which (according to Di Gregorio, too) he and Gegenbaur rejected. Actually, Haeckel's point was to explain away the type concept by reducing it to an effect of heredity from a common ancestor, which conserved the common features of the group. The type was no longer a cause of form, but a result of other Darwinian processes.

Haeckel's point (II) attributed departures from the ancestral form or type to the effects of natural selection and of adaptation to the conditions of existence. It then continued, along the same lines as point (I), to reduce unity of type to the product of a common history of adaptive changes, as preserved by heredity. Adaptation and its own result, unity of type, interact at the end of the quote to account for all the diverse forms of life.

Di Gregorio twists the quote to make it conform to the standard historiography. He turns point (II) completely on its head, to make it give priority to the law of unity of type instead of to natural selection. In his translation of the passage, unity of type is the law that can "encompass the inheritance of earlier adaptations," and with it, "mould the multiplicity of organic forms." Type, rather than evolutionary history, is made into the ultimate cause of form, a gross distortion of Haeckel's historical thinking.

Haeckel actually made unity of type into a historically contingent result of the interplay of adaptation and heredity. The taxonomic types were not autonomous causes of form; neither were they predetermined at the Creation, or by transcendental ideas; nor did they result entirely from functional necessities. This early notebook entry offered a preview of the balance Haeckel would strike between the conservative influence of heredity, which kept the type more or less unified, and the progressive influence of adaptation, which kept modifying types and creating new ones. Haeckel's detractors insist on reading only one or the other side of the balance, usually the conservative side that maintained unity of type.

Haeckel's manner of expressing Darwin's ideas in his notebook in terms of paired, interacting laws was certainly reminiscent of Bronn. So is the fact that he made adaptation the higher law, which somehow entailed or embraced unity of type. In fact, the quote was a close paraphrase of a passage of Bronn's, but not from his early theoretical works. It was from his Darwin translation, near the end of chapter 6 of the *Origin*, entitled "Difficulties on Theory." But the talk of law was not entirely Bronn's invention. In the original, Darwin himself wrote that "all organic beings have been formed on two great laws—Unity of Type, and the Conditions of Existence."[17] Bronn took the small liberty of promoting Darwin's laws into Bronnian *Grundgesetze*, and making the second one into "*Adaptation* to Conditions of Existence," which corresponded better to his own *Grundgesetz* of adaptation. Darwin's claim that "unity of type is explained by unity of descent" came through pretty much intact in Haeckel's note, as did Darwin's claim that "the law of Conditions of Existence is the higher law; as it includes, through the inheritance of former adaptations, that of Unity of Type."[18] To this point, Haeckel's passage was a reasonable rendition, albeit in Bronn's terminology, of what Darwin actually wrote, and there is no reason to conclude from it that Haeckel misunderstood it or failed to free himself from Gegenbaur's, or even older, ideas about types.

All that was really new—or perhaps I should say, old—in Haeckel's note was the reference to the production of *Mannigfaltigkeit* by the interactions between the laws of type and adaptation. Darwin did not say anything about this on the corresponding page of the *Origin*. Of course, Darwin would not have disagreed with the basic notion that new forms received many features not only of the ancestor/type through heredity, but also many new and adaptive ones that had been accumulated by natural selection, and hence that the two "laws" acted together. This was essentially what Darwin described in the portion of the chapter illustrating evolutionary transitions. He depicted the flying squirrels, for example, evolving various sorts of skin flaps between front and rear limbs to use as parachutes.[19] Through all the changes, they remained clearly recognizable as squirrels. They did not sprout wings out of nowhere or turn into birds in one evolutionary leap. Natural selection was always depicted as having to work with the resources provided by the ancestor/type and by gradually accumulating variations.

Still, Darwin did not express himself in quite the same way as Haeckel, and the words did make a difference. The language of *Mannichfaltigkeit* and of interacting laws brings to mind an old problem of German morphology, one that was once addressed by Meckel and later by Bronn. The note suggests that Haeckel was interpreting Darwin with an eye toward addressing German morphologists in their own language and solving problems of theirs that Darwin had not addressed explicitly. Haeckel was finding opportunities for himself to break new ground in Darwinian morphology, and he began to capitalize on them in 1863, when he addressed the *GDNÄ* meeting.

The Stettin Presentation

Haeckel's address at the Stettin meeting, entitled *"Über die Entwicklungstheorie Darwins"* (On Darwin's theory of development), struck a masterful balance between the revolutionary and the traditional. It proclaimed the arrival of Darwinism as a theory and a worldview, but also made overtures to the more traditional morphologists and paid due respect to the departed Bronn. The title of Haeckel's paper and his continual reference to Darwin's theory as a "theory of development" (*Entwicklungstheorie*) made it sound like an extension of older approaches. He might easily have called it a "theory of descent" (*Descendenz* or *Abstammung*), "selection," "transformation," or even "metamorphosis," instead. (The word "evolution" was not yet commonly used in its modern sense, either in German or in English.)

Haeckel called for German biologists to unite behind the new theory, and he presented it not merely as an intellectual matter, but as a necessary step toward German cultural and political unification and a means of affirming belief in progress. The venue was well chosen for such purposes. The *GDNÄ* meetings provided the largest and broadest scientific audiences available, and they had been a symbol of German aspirations for cultural and political unity and progress ever since their inception in 1822. They were the brain-child of Haeckel's predecessor at the University of Jena, the zoologist and liberal-nationalist political activist Lorenz Oken. Oken conceived the *GDNÄ* as a national society, not bound to any one German state, and it brought together scientists from all over the German-speaking world at a time when the Metternich system discouraged any efforts toward German unification. He also implemented egalitarian and democratic principles in its internal organization and membership policies, in contrast to the Leopoldine Academy and older scientific organizations. As a result the *GDNÄ* is generally viewed by historians as an early manifestation of German liberal political aspirations.[20]

In Stettin, Haeckel got to the political message right away. He described Darwin's theory as one that, "on the one hand, threatens to shake the foundations of an entire, great, scientific theoretical structure, which for centuries enjoyed, and continues to enjoy, almost universal recognition; and on the other hand seems to intrude in the deepest way into the personal, scientific, and social views of every individual."[21] That the theory entailed a change in worldview, he said, was apparent from its basic premise, which he identified as universal common descent: "All the different animals and plants that still live today, as well as all organisms that have ever lived on earth, were not, as we have been used to assuming from our early youth on, independently created, each one by itself, in its species. Rather, despite their extraordinary diversity [*Mannigfaltigkeit*] and difference, they have developed gradually over the course of many millions of years from just a few, perhaps even from *one single ancestral form*, a most simple proto-organism"[22] (emphasis in original). Consequently, he continued, humans would have to seek their ancestry in apelike mammals and, more distantly, in primitive marsupials, reptiles, and fish.[23]

Haeckel postponed discussion of the mechanism of evolution until later in his talk. In this initial formulation, he focused on the fact that evolution occurred somehow and that all living creatures were related by descent. He was trying to cast his net as broadly as possible and to unite as many biologists as he could behind the general idea of evolution and against religious dogma and political conservatism. He argued that the idea of common descent was not new, but had long been recognized by a whole pantheon of great naturalists. Haeckel contrived to make Goethe, Oken, Lamarck, and Geoffroy all into champions of common descent. By his reading even the staunch antievolutionist Cuvier was a forerunner of Darwin. Cuvier had argued that species were always so perfectly and delicately adapted to their conditions of life that they could not possibly change, but to Haeckel, this implied that he recognized the overriding significance of adaptation and functionality as explanations of form and prerequisites for survival.[24]

Historians have pointed out that Haeckel was mistaken about his predecessors and lamented that after fifty years or more of repetition, his misattributions came to be commonly accepted.[25] But they overlook the method in his "mistakes" and the rhetorical advantages they brought him. In Germany it was always good to have Goethe on your side in an argument, and the poet helped fend off the objection that struggle and adaptation deprived the natural world of its beauty. At GDNÄ meetings, Oken was an important icon, too. Bringing him into the evolutionary fold reinforced the connections among Darwinism, progress, and German liberalism and might have helped to sway the morphologists in the audience who had learned their zoology from Oken and his books. Haeckel's twisted reading of Cuvier preempted an important counterargument against species transformation, and by stripping Lamarck, the one bona fide evolutionist of the lot, of his most objectionable features, Haeckel subsumed a potential competitor into his system. Haeckel's Lamarck never invoked subtle fluids, an inherent drive to perfection, or any role for the animal psyche, but only the purely mechanistic effects of use and disuse and of external, environmental forces.

Haeckel had all the great naturalists in his camp, and he recognized only one other camp his listeners could possibly join in the growing conflict over evolution: "On the flag of the progressive Darwinists stand the words: '*Evolution and Progress!*' From the camp of the conservative opponents of Darwin sounds the cry of: '*Creation and Species!*'"[26] (emphasis in original). Neutrality in the conflict was not an option. The differences between the two sides were growing daily and soon everyone would have to choose between them: "Every day wider circles are caught up in the powerful movement. Those who stand aloof will also be pulled into the maelstrom, and for good or ill even those who would like to stand above the fray will have to favor more the one party or more the other after all."[27]

Haeckel then began to present the case for the Darwinian side, continually comparing Darwin's theory with earlier and more familiar theories, emphasizing their common ground and their common rejection of the biblical account. He broadened the definition of Darwinism to subsume all possible naturalistic theories of life and alternatives to biblical creation: "If we put all previous creation theories together, we can organize them, one and all, into two opposite rows: the one row of cosmogonies maintains, with the Mosaic creation story, that all species of living things were called into being independently, each by itself, by the will of an omnipotent Creator; the other, that they are all branches of a single tree and products of one and the same, constantly working, natural law of progressive development."[28]

For the time being, following the letter of the *Origin of Species*, he left open a loophole for belief in divine creation of the first species, combined with evolution of the rest: "Thus no species, *maybe* with the exception of the first, was individually created. Much rather, they all came forth, over the course of immeasurable lengths of time, from just a few or from one, single, *maybe* spontaneously generated original form"[29] (emphasis added).

As I argue below, Haeckel's *Generelle Morphologie* of 1866 closed all such loopholes for the creationists and teleologists. When the Stettin lecture was reprinted in 1878, Haeckel saw to it that the "maybes" in the above quote were edited out, in keeping with his more restrictive interpretation: "Thus no species, *not even* with the exception of the first, was individually created. Much rather, they all came forth . . . from just a few or from one, single, simplest, spontaneously generated original form"[30] (emphasis added).

Once he had made his case for the general features of evolution, species change, and common descent, Haeckel began to deal with the more controversial and specifically Darwinian elements: the interplay of heredity and variation, which both preserved ancestral characteristics and introduced new ones; the inevitability of Malthusian increase, overpopulation, and struggle for life; and the mass destruction of life that must follow. Haeckel explained that because the destruction was selective, nature did, in effect, what practical breeders did, and Haeckel referred to the process as *natürliche Auslese* or *natürliche Zuchtwahl* (sometimes also *Züchtung*, as Bronn had translated it). The result was continual progress and perfection, but not in one straight line. There was divergence, with lineages radiating in all directions from their common ancestors.[31]

Haeckel emphasized that human origins were to be included in Darwin's scheme, and that not only the human form, but human culture had evolved out of lower incarnations: "Considering all that we know of the earliest periods of human existence on the earth, we are entitled to assume that the human, too, neither sprang forth as a fully armed Minerva from the head of Jupiter, nor came out of the hand of the Creator as a fully grown and sin-free Adam. Rather, he worked his way up, extremely slowly and gradually, from the primitive condition of animal rawness to the first simple beginnings of culture."[32]

On the question of variation and its causes, Haeckel hewed closer to Darwin here than in his later works. He took Darwin's pluralistic view and allowed for many causes of variation. Variations could be inborn, or acquired during later life in the "Lamarckian" manner, as the result of environmental influences, including those emanating from the biological environment. Variations, he wrote, "are in part already present in the germ of the individual, in the egg. . . . In part, the individual characteristics are only acquired during the life of the individual, through adaptation to the external conditions of life, especially through the interrelationships that every organism has with all the others around it."[33]

After giving an account of variation and selection according to Darwin, Haeckel returned to the main theme of the presentation: progress. He argued that the result of natural selection would have to be *progressive development*. The words were repeated and emphasized throughout the text. At first, Haeckel stated that progress could be expected as a general result of natural selection "on the whole" (*im Großen und Ganzen*), as opposed to making progress a necessity or law of nature. Before long, however, he did indeed refer to it as a law of nature that applied not only to plants and animals, but to humans and their political affairs as well: "Further, we find the same *law of progress* in effect everywhere in the historical development of the human race. Quite naturally! For here, too, in civil and social conditions, the same principles of struggle for existence and natural selection are what drive nations [*Völker*] irresistibly forward and raise them stepwise to higher culture" (emphasis in original).[34]

In a similar vein, Haeckel went on passionately to declare that all the efforts of benighted priests and politicians to stifle progress were doomed to failure, because life would always bound forward once the temporary obstacles fell away. The law of progress almost sounded like a cause of historical change, as it had been to Bronn. At the end of his talk, Haeckel quoted Bronn at length on Darwin's potential to unify all the phenomena of natural history and eventually overcome all the remaining objections. He relied upon Bronn's authority again during a discussion period, following the talk, when he was challenged by a geologist on the paleontological evidence for steady progress.[35]

Over the next three years, however, Haeckel went on to build his own complete, unified system of the organic world, based on a narrower interpretation of Darwin than he had offered at Stettin. He toned down the deference to Bronn and the overtures to developmental interpretations of evolution, and he set forth his own distinct version of Darwinism.

The Hardening of Haeckelism

Haeckel's introduction to Darwinism in 1860 had come at the end of a time of soul-searching, when he finally committed himself to a career in science and abandoned the religion of his youth. While still in Italy, where he had barely recovered from the shock of Johannes Müller's death, he was prompted by the death of a young friend to write to his beloved Anna Sethe about a crisis of religious faith:

Lachmann is now the second of my little circle of close friends and fellow students whom senseless fate, in the course of a single year, has torn away from his lively sphere of activity. . . .

How anyone can console himself in such cases with the idea of an "omni-benevolent, wise, loving Providence" is really entirely unclear to me. I, at least, am fully unable to grasp the idea of a Providence, a personally reigning God, who gives His creatures the most noble gifts, only to let them make as little use of them as possible, and who sets them in the loveliest fullness of domestic happiness, only to tear them away most cruelly and violently.[36]

With eerie prescience, he worried in the same letter about losing Anna at a young age, too, and having to survive without her and without the consolation of religion. He predicted that he would not be able to go on with his life.[37]

On February 16, 1864, Haeckel's thirtieth birthday, his worst fear was realized. Anna died suddenly, apparently of some kind of abdominal infection. Haeckel reacted much as he had predicted in the letter to Anna several years before. He was quite incapacitated by it, too upset even to attend the funeral. He contemplated suicide, but eventually was able to find comfort, or at least distraction, in Darwinism. Darwinism showed him how there could still be progress, beauty, and harmony in a world full of struggle and senseless death.

His biographers generally treat Anna's death as a pivotal moment in Haeckel's personal and intellectual development. Richards is drawing renewed attention to it as the event that made Haeckel's career, that gave him his characteristic convert's zeal to promote Darwinism and to challenge God, the clerics, and dogma in all its guises. Di Gregorio, too, sees the memory of Anna as staying with Haeckel and inspiring his most creative work for decades to come.[38]

As Haeckel told the story himself, he threw himself into the project of making just one major scientific statement before, he had thought, ending his life. The result was the two-volume *Generelle Morphologie*: "At the time I didn't think I could overcome that blow. I thought my life was over and done with, and only wanted to epitomize, in one last, lengthy work, all the new ideas that Darwin's newly flourishing theory of evolution had inspired in me. And so, *Generelle Morphologie* came into being, amid hard struggle. It was written and printed in less than a year. I lived then quite like a hermit, allowed myself barely 3–4 hours sleep daily, and worked all day and half the night. I lived in such strict asceticism that I must really be amazed to stand alive and healthy before you today."[39]

Certainly, the loss had a lasting effect on Haeckel, but his conversion to Darwinism was already well prepared and well under way beforehand. Haeckel had lost his religious faith by the late 1850s, following the premature deaths of friends and colleagues. He had been a leading Darwinian since his Stettin talk of 1863. And there was no shortage of vitriolic outbursts against the Catholic Church, dogma, and superstition at Stettin and in earlier correspondence. What was new after Anna's death was perhaps an increase in the zeal and emotion behind the Darwinism, but also a certain "hardening"[40] that could be detected in his theoretical positions as well as in his attitude toward his professional position. In a letter to his parents a year after his loss, Haeckel said he felt the experience had "hardened" him personally and that he had become more resolute, less sensitive to criticism.

Yes, in this terrible year, a man has grown out of the child, and the daily repeated hammer blows of unavoidable fate have forged the steely character in me that had been missing so much before. How often must I now think of the saga of the hard-nosed [hartgeschmiedet] count, which my Anna read to me so often when I was too soft and indecisive. "Get hard!" she used to conclude with a little, teasing joke.[41]

This hardening of Haeckelism is evident in the contrast between Haeckel's 1863 lecture and his 1866 book. From 1866 on, Haeckel defined his Darwinism more strictly, made his break with developmentalism and teleology more explicit and thorough, focused on countering Bronn's objections to Darwin's theory, and developed a thoroughgoing evolutionary system of morphology. Whereas his 1863 lecture had reached out to all morphologists and been agnostic about the origin of the first species, the later Haeckel tolerated not a hint of any predetermined, providential, or self-directed life-processes. The biological world had to be an extension of the physical, devoid of any independent purpose or agency. Darwinian evolution proceeded according to discoverable laws, perhaps, but those laws were blind and purposeless, like the laws of physics. And there was no part of biology—not even human biology, anthropology, psychology, or the humanities—that escaped Haeckel's reduction to mechanistic, monistic principles.

By Haeckel's own estimation, *Generelle Morphologie* was the project "that determined the programmatic direction and method of my entire remaining life's work."[42] Indeed, virtually all of his articles and monographs referred back to the authoritative statements in *Generelle Morphologie* upon which they elaborated. The same was true of most chapters in his popularizing *Natürliche Schöpfungs-Geschichte* (Natural history of creation, or, more literally, Natural creation story or history). No matter how many times he revised it, it always deferred to the authority of *Generelle Morphologie*.[43]

Banishing Teleology: Biology as a Mechanistic Science

Haeckel's *Generelle Morphologie*, as its subtitle declared, was a treatment of organic forms that was placed "on a mechanical basis through the theory of descent, as reformed by Charles Darwin." The dedication to Gegenbaur reiterated that *Generelle Morphologie* was intended above all to be a complete system of "mechanical morphology," and the mechanistic outlook was emphasized throughout the book. In the foreword, Haeckel explained that within biology, only the subdiscipline of physiology had matured enough to join the ranks of the mechanistic sciences and give up the old "dualistic" reliance on both physical and nonphysical (e.g., mental, spiritual, transcendental, or divine) causes and explanations. Morphology was far behind; it still needed to be cleansed of its vitalistic and teleological dogmas and belief in miracles, and reduced to a "monistic," mechanistic approach. Darwin had pointed the way by providing a mechanistic explanation of evolution, and Haeckel was carrying on from there and creating a mechanistic, evolutionary morphology.[44] There was much more to be done than Darwin could ever have accomplished in his single chapter on morphology.

As in all of Haeckel's forays into the history of biology, throughout *Generelle Morphologie* he gave Lamarck and Goethe priority for the idea of species transformation and common descent. However, Darwin was at the top of the trinity of great evolutionists, because of the mechanism of natural selection. Echoing Bronn, Haeckel found that struggle and selection were what made evolution *necessary*. Divine intervention, teleology, dogma, and superstition were now superfluous: "Only Darwin's overpowering arguments were able to shoot open a breach in the castle walls of belief in miracles, a decisive breach that opened the way for the unbeatable thoughts of the calculating, synthetic mind to enter the innermost refuge of vitalistic foolishness."[45] Haeckel expected Darwin's theory to rid biology of every form of miracle, from overt Creation myths and fixity of species to pseudo-scientific forms of vitalism, teleology, organizing principles, or creative forces: "*We see in Darwin's discovery of natural selection in the struggle for existence the most striking evidence for the exclusive validity of mechanically operating causes in the entire field of biology. We see in it the definitive death of all teleological and vitalistic interpretations of organisms*"[46] (emphasis in original).

However, Haeckel was not satisfied with the way Darwin's principles had been applied so far. There were some soft spots that needed to be hardened in order to disallow every last remnant of teleology and divine providence and to disarm Bronn's criticisms. One was Darwin's indecisiveness about where the first organisms came from, a point upon which, Haeckel, too, had wavered in 1863. Now, in his special chapter on descent and selection in *Generelle Morphologie*, Haeckel began with the clear assertion that the very first organisms arose through physical and chemical processes, driven by the same laws of nature that applied in the nonliving world. In other words, it was what Haeckel considered a "mechanistic" process.

If every subsequent step of the evolutionary process was mechanistic in the same sense, then the organic world was only an extension of the physical and could be approached using the same scientific methods: "We do not think we can emphasize this extremely important point enough. It is the unassailable citadel of scientific biology. As long as one keeps this fundamental point in mind, one will never underestimate the immeasurable significance of the theory of descent."[47] There was no special *biological* realm of purpose and design, and no exception for human origins or human culture. That the cosmos as a whole might still exhibit plan and purpose was left as a possibility, however.

Balancing Adaptation and Heredity, Creativity and Constraint

A weak link in the chain of mechanistic steps was heredity. As an internal, biological process, heredity was strictly conservative, according to Haeckel. By definition, it could not generate anything new and purposeful, but could only repeat in the offspring what had already been present in the parent. Variation, which did have a good chance of yielding something new and purposeful, was a separate process, triggered not within the organism but by independent and unpredictable physical forces in the environment. That way, there was no unbroken chain of internal, biological causes going back to the origin of life, and no way to argue that evolution was a simple expression of a potential that was already present in the first created organisms.

The creative part of Haeckel's mechanism of evolution, therefore, was outside the organism. The physical environment was continually modifying organisms, generally—but not always—in an "adaptive" or "progressive" direction. Once the environment acted, its effects could then be preserved by the internal, conservative principle of heredity, if the modified organism survived in the struggle for life: "Every organism transmits to its offspring not only the morphological and physiological characteristics that it inherited from its parents, but also a portion of those characteristics that it acquired itself, during its individual existence, through adaptation."[48]

Over and over, Haeckel emphasized that the process of acquiring a new characteristic was a purely mechanistic response to the environment, capable of branching off in any direction. The environmental forces that impinged upon the individual were so many and varied that there was no way to set a limit upon their actions and effects. Moreover, no two individuals would experience the same environmental forces in exactly the same way and at exactly the same time and place, so even in an apparently uniform environment, variation was inevitable and unpredictable.[49]

This combination of conservative heredity with ubiquitous, unpredictable, and unlimited variation constituted Haeckel's solution to Bronn's problem of species boundaries. Nature had no special "lawlike means" to keep returning varieties to the species norm, following anomalous divergences. Variation was a purely mechanistic response to environmental cues, and heredity passed all variations, old and new, on to the offspring. There was no provision for preferentially getting rid of recent or excessive variations, except by natural selection.

The mechanistic and unpredictable nature of variation was difficult to reconcile with Haeckel's belief that it would generally be progressive. He clearly expected organisms to vary in constructive ways most of the time, and he tended to use the terms "variation," "adaptation," and "progressive heredity" interchangeably. Yet his enthusiasm for progress was at least matched by his abhorrence for teleology and for theories of directed evolution. He thought he could have progress without teleology as long as the direction of change was not dictated by a law or transcendental plan, or by anything within the organism like an idea or soul, and changes were not guaranteed to be an adaptation. Variation, therefore, was not "random" in the sense of being equally likely to go in all directions, but it was still undesigned and unguided by any sort of mind or purpose. Variation was also gradual and cumulative and did not yield complete adaptations all at once. Mindless, purposeless natural selection still had to accumulate and mold favorable variations into adaptations and give evolution its generally progressive direction.

Although Haeckel tended to play down the possibility, he did write specifically in *Generelle Morphologie* that not every environmentally induced change would automatically be in a positive direction. There was also a provision for "indirect adaptation," in which the environmental effects did not produce any immediate changes at all, positive or negative. The changes only became manifest in the offspring of affected individuals. This corresponded to Darwin's "individual differences" caused by the environmental disturbances or destabilization of the reproductive system,[50] and, incidentally, allowed Haeckel to count a new trait as "acquired" even if the individual was born with it. The most important effect of environmental action, therefore, was variation and not adaptation: "The most general expression of the *Fundamental Law of Adaptation [Grundgesetz der Anpassung]* may be found in this statement: '*No organic individual remains absolutely the same as the others'*"[51] (emphasis in original). This emphasis on individual variability strongly distinguishes Haeckel's theory from the "Lamarckism" that is usually attributed to him.

Note also Haeckel's appropriation of Bronn's language for new purposes. Adaptation was still a *Grundgesetz*, as Bronn had it, but now it governed the production of new and generally beneficial variations, rather than the elimination of maladapted species. It did not appear to have the same epistemological status as Bronn's version, however. There was no claim that it held *necessarily*, as a matter of logic as well as empirical observation.

While hardening his positions (and Darwin's) on the origin of life, variation, and heredity against teleological and developmental interpretations, Haeckel continued to appropriate developmental language and metaphors for his own purposes. Critics might have been able to accuse Darwin of dispensing, unscientifically, with the rule of law in evolution, but Haeckel gave them laws in great abundance. On closer examination, however, Haeckel's laws did not determine the precise historical succession of forms. The laws all applied to individual components of the evolutionary process, and there were so many of them, often working at cross-purposes, like heredity and variation, that one could never use them to predict where evolution would lead. In his approach to biological law, Haeckel had taken a page from Bronn's book in order to generate unpredictable and diverse results from a system of ostensibly deterministic laws.

For example, law number one of heredity was the "Law of uninterrupted or continuous heredity" (*Gesetz der ununterbrochenen oder continuirlichen Vererbung*), which mandated the resemblance between parent and offspring: "In most organisms, all immediately successive generations are either nearly identical to one another in all morphological and physiological characteristics, or at least very similar." Meanwhile, law number six, the "Law of adapted or acquired heredity" (*Gesetz der angepassten oder erworbenen Vererbung*) allowed for differences as well: "All characteristics that the organism acquires during its individual existence through adaptation, and which its ancestors did not possess, can be transmitted by the organism to its offspring, under favorable circumstances."[52] Almost every law had such a counterlaw, and the resulting system could be bewildering to the uninitiated, but it contained a ready explanation for every imaginable evolutionary outcome. It was a way of translating Darwin's practice of constructing many ad hoc case histories into a language of law and determinism.

Recapitulationism Reconsidered

Haeckel also took over the traditional embryological metaphors for evolution, while eliminating their teleology and determinism. He did not deny the apparent parallelism between the development of the individual embryo and the forms of lower animals, either in the fossil record or the idealized scale of nature, but he gave it a strictly evolutionary interpretation and revised the standard terminology. The term *Entwicklung* had been used before for any progressive morphological sequence, whether in the embryo itself or in the fossil record, and in the fossil record it might have applied to a process of linear transformation, or to a branching tree of common descent, or even to a scheme of successive spontaneous generations like Bronn's. Haeckel now distinguished among these meanings of *Entwicklung*. He called individual embryonic development "ontogeny" (*Ontogenie*), and the historical development of the species, "phylogeny" (*Phylogenie*). He also specified that any parallelism was only between those two sequences, not with a Bronnian succession of species, and certainly not with an idealized scale of forms. Parallels were not due to any transcendental plan or developmental law that determined both sequences. And, most important, internally generated changes or trends in ontogeny did not cause evolutionary change or determine its direction. On the contrary, changes in ontogeny were caused by historical events— interactions between organism and environment—that produced and preserved successful variations. The twists and turns of ontogeny were caused by the history of the species, and, properly interpreted, they could, in principle, yield an actual record, or recapitulation, of phylogeny.[53]

The main reason ontogeny was able to recapitulate phylogeny was Haeckel's law of heredity number nine, the "Law of homochronic heredity": "All organisms can transmit the particular changes that they acquired through adaptation at any time in their individual existences . . . to their offspring at exactly the same time of their lives."[54] In other words, heredity preserved not only the sum of the parental characteristics, but also the sequence in which the parent expressed or acquired them.

Two points are important to note about this law of homochronic heredity. The first is that it was entirely unoriginal, but came from Darwin, by way of Bronn. Darwin had made the same generalization and even suggested its applicability to embryology. Darwin wrote, in the chapter on variation under domestication, that, "at whatever period of life a peculiarity first appears, it tends to appear in the offspring at a corresponding age, though sometimes earlier. . . . I believe this rule to be of the highest importance in explaining the laws of embryology." In translation, Bronn gave the passage a slightly stronger wording, making the peculiarity *always* appear at the corresponding age, though sometimes earlier.[55] And although Haeckel followed Bronn and kept the timing exact in his formulation of the law, he introduced other laws that allowed variations to be moved up to earlier developmental stages, or sometimes even to be delayed to later stages.

Whenever new characteristics were acquired somewhere toward the end of individual development, the earlier portion of development was left unaltered as a record of past acquisitions, and ontogeny provided a usable recapitulation of the earlier portion of phylogeny. When new characteristics happened to get inserted in the middle of development, the historical record of the affected organ or part was "falsified" in some way from the insertion-point on. Other eventualities could obscure the phylogenetic information as well. For example, further laws of heredity provided for the recapitulational process to take shortcuts that hastened the appearance of important adaptive characteristic or made the whole process faster and more efficient.[56] Ontogeny could therefore never be read like a book as a continuous record of phylogeny. The job of the evolutionary morphologist was to exercise judgment and identify the usable snippets that contained unfalsified historical information.

The second important point about Haeckel's law of homochronic heredity is that it did not commit Haeckel to the unrealistic claims about "terminal addition" and preservation of ancestral adult forms that Gould attributes to him. According to Gould, Haeckel's version of recapitulationism could work only if new characteristics were acquired (by the adult) through use and disuse or environmental effects, and incorporated into the ontogenetic sequence at the very end. Logically, therefore, Gould argues, any recapitulationist must be committed to acquired characteristics and terminal addition.[57] But the correctness of Gould's claim depends entirely on what it meant for recapitulationism to "work." If ontogeny were expected to supply a clear and continuous recording of phylogeny, then, indeed, it could work only Gould's way, which is to say that, in practice, it did not work at all, and Haeckel would appear to have been very naive. But perhaps Haeckel made recapitulation work in some other way.

Among Haeckel's many neologisms, none corresponded to "terminal addition," and among Haeckel's many laws, none tacked on variation at the end of development. Gould's interpretation is based only on the slogan that ontogeny recapitulates phylogeny. Instead of requiring terminal addition, Haeckel applied the law of homochronic heredity, which allowed new variations to be inserted *anywhere* in the sequence of developmental stages but had them maintain their relative positions. When reconstructing phylogenies, Haeckel might well have wished for more cases of terminal addition, but he knew they were not the rule, and his methods were designed to take that into account. Indeed, Haeckel provided a thorough analysis of the many things that could go wrong and falsify the story of phylogeny. Not only could new variations appear at any time during ontogeny, but their relative position in the sequence could be altered as well, despite the law of homochronic heredity. Anticipating Gould, Haeckel even discussed the various forms of "heterochrony," in which the appearance of a particular trait is accelerated or delayed relative to other traits.

Haeckel also required a mechanism for continually accelerating the whole developmental process to compensate for the accumulating insertions of new variations that might otherwise keep lengthening the gestation period. This overall acceleration was provided for by the "Law of abbreviated or simplified inheritance" that allowed ontogeny to take shortcuts and skip historical stages: "The chain of inherited characteristics that are transmitted in a particular order . . . becomes shortened over the course of time, in that individual links in that chain are dropped."[58]

As a result of all these processes, recapitulation never really was expected to work in Gould's sense at all. As Haeckel noted himself, "Even though, on the whole, the individual developmental history of each organic individual is a brief repetition of the long paleontological development of its ancestors—ontogeny is a brief recapitulation of phylogeny—even so, we must nonetheless add, as a very important elaboration of this fundamental proposition, that this repetition is never quite a complete one."[59] Haeckel's system of recapitulation and falsification nonetheless worked for Haeckel—as an interpretive lens through which to view the embryo and make inferences about its evolutionary history.

Admittedly, Haeckel did have a penchant for sloganeering, and he often made dogmatic and simplistic assertions about the "fundamental biogenetic law" (*biogenetisches Grundgesetz*) that ontogeny recapitulated phylogeny. But in the fine print, and in practice, the matter was never so simple. The biogenetic law was fundamental in the sense that it was the background assumption, or null hypothesis, with which the interpretation of ontogeny was supposed to begin. Environmental conditions, by triggering the other laws of heredity and variation, virtually guaranteed that the underlying recapitulational story would be falsified to some extent—perhaps to a great extent. But the analytical framework still applied even if no recapitulation could be detected at all. The falsifications were parts of the evolutionary stories that Haeckel told.

In the larger history of German morphology, then, Haeckel's recapitulationism was not a simple return to older models such as Meckel's. To the extent that the older recapitulationists accepted evolutionary change at all, they considered the sequences of ancestral forms and of embryonic forms to be independent, parallel products of common, underlying laws or plans. In Haeckel's system, there was no underlying plan, and the causal arrows pointed from phylogenetic history to ontogeny. During the history of a lineage, external forces changed the organism, and the change was merely recorded in heredity, to be repeated during future ontogenies. The reason the human embryo went through a stage with gill arches and gill pouches, resembling a fish's gill slits, for example, was not a law requiring the developmental process to pass through a fish stage, or even a fishlike stage. It was rather the historical fact that humans happened to share an ancestor with fish, an ancestor that had something like those structures. As external forces, together with natural selection, shaped the lineage gradually into humans, they did so mostly by tinkering with later stages and other organs in such a way as to leave the gill-arches recognizable.

The phenomenon of recapitulation had tremendous practical significance for Haeckel, because it could be used in reconstructing the history of the lineage and its taxonomic relationships. It provided an answer to Bronn's complaints about Darwin's many hypothetical historical narratives that could not be supported by the fossil record. The availability of phylogenetic information in every embryo gave the evolutionary morphologist new authority to describe the history of life, which had been the province of the paleontologist.

Recapitulation also worked for Haeckel in the sense that no matter how much the record of any one lineage was falsified by adaptive changes, ontogeny always provided support for the general principles of Darwinian evolution. There was no way to explain ontogeny's illogical twists and turns without reference to unpredictable historical events, caused by physical and chemical processes in the environment, which modified inherited tendencies. Conversely, there was nothing in ontogeny that required any explanation other than heredity and adaptation.

No separate, "dualistic," nonmaterial realm of ideas, souls, and purposes was needed to explain anything.

This point was central to Haeckel's system of thought, and he repeated it throughout *Generelle Morphologie* and his later writings. Even his detailed monographs on the description and classification of various groups of organisms reiterated that ontogeny and phylogeny were ultimately mechanistic, "monistic" processes. For example, he concluded his account of the calcareous sponges with a little bit of "sponge philosophy":

All phenomena that appear in the morphology of the calcareous sponges can be explained completely by the interactions of two physiological functions, *heredity* and *adaptation*, and we need no other causes to make their appearance understandable. All causes that reveal themselves actively at all in the morphology and physiology of the calcareous sponges are *unconscious, mechanistic causes (causae efficientes)*; nowhere do we encounter *conscious, purposeful causes (causae finales)*. We can perceive the rule of unchangeable laws of nature everywhere, the effects of a premeditated plan of Creation nowhere.[60] (Emphasis in original)

The ontogeny and phylogeny of the sponges therefore provided empirical support for monism.

Haeckel was not content with banishing teleology and divine providence only from the study of ontogeny and phylogeny. He had ambitions for unifying all of biology within his monistic, mechanistic scheme by incorporating all the biological disciplines into the grand project of reconstructing evolutionary history and seeking historical explanations. He made his intentions clear in his inaugural address at Jena, upon becoming a full member of the Philosophical Faculty in 1869.[61]

Haeckel argued that zoology was not yet a unified, coherent science in the nineteenth century. Anatomy and physiology had developed out of medicine, and paleontology out of geology. Psychology was still mainly part of what he called "speculative philosophy," and traditional natural history or systematic zoology had steered its own course and taken no notice of what was going on in the other areas. None of the biological disciplines, least of all systematic zoology, had yet concerned itself with historical, phylogenetic, and above all mechanistic explanations of its subject matter. Haeckel planned to change all that by building systematic zoology into a comprehensive, total science, or *Gesamtwissenschaft,* that would incorporate all the other disciplines into its evolutionary framework. Even psychology or any part of philosophy that dealt with mental phenomena would be included, because, under monism, the mind was a product of the evolution of the nervous system. The new *Gesamtwissenschaft* would draw on findings from all its subordinate fields to reconstruct phylogenetic history, and in return it would provide proper, mechanistic explanations of all biological phenomena: "And so, from the little, despised seed of zoology a tree of science will develop, which in the future will bring all the other sciences into its shade and out of whose roots they will all more or less have to draw their nourishment."[62]

Haeckel's total system of zoology was hard and uncompromising on the principles of external, mechanistic causes and historical explanation, but remarkably flexible otherwise. Haeckel was frank about the practical limitations of using ontogeny as a source of evidence about phylogeny and acknowledged that alternative readings of the record were possible. Indeed, his laws of heredity and development had enough loopholes to accommodate any number of individual cases of "falsified" recapitulation.

Much later, Haeckel was also willing to back off from his rhetoric in *Generelle Morphologie* that seemed to equate variation with adaptation and progress. In later works he dropped the terms "progressive" or "adaptive heredity," which had referred to the inheritance of any environmentally induced modification, in favor of the more neutral "transformative heredity." The category had included regressive as well as progressive changes all along, Haeckel claimed, but he finally decided that the original terminology overemphasized the positive.[63]

The concept of "indirect" adaptation or transformation also gave Haeckel some leeway for eventually accommodating evidence for genetic mutations, while still insisting that all changes were acquired in response to environmental effects. For example, if an albino suddenly were born in a purebred colony of black mice, he could count albinism as an acquired characteristic. By Haeckel's interpretation, it had not previously been part of the mouse's heritage, so it must have been acquired when some unidentified, external cause altered the parent in such a way as to make it produce the albino.[64]

Darwinizing Baer

Haeckel was also flexible on the precise relationship between ontogeny and phylogeny. Before he could complete his assimilation of embryology into a Darwinian *Gesamtwissenschaft* in the 1860s and 1870s, Haeckel still had to answer the old arguments of Karl Ernst von Baer against species transformation and especially against drawing simple parallels between individual and species development.

Biologists from E. S. Russell to Gavin de Beer and Stephen J. Gould have held up Baer's theory of differentiation from the general to the special as a far superior account of recapitulation (or the appearance thereof) to Haeckel's. Gould, along with Bowler and several other authors, makes Baer into the more modern thinker, whose view of development is easily brought into harmony with Darwin's, while relegating Haeckel to the ranks of the transcendentalists and teleologists. They wonder how anybody could have accepted Haeckel's recapitulationism after Baer's critique.[65]

It is a strange question for a modern Darwinian to ask, considering that Baer wrote his critique from a teleological, transcendentalist perspective. He took for granted that any *Entwicklung* had to be guided by the idea of the final form, and that species transformation, if it occurred at all, would have to be a form of *Entwicklung*, entailing differentiation within an archetype. It is also an anachronistic way of looking at Baer, who aimed his critique at the recapitulationism and transformationism of his day, not at Haeckel.

I suspect that Baer's admirers do not really have the original Baer in mind at all, but a cleaned-up, "Darwinized" version, without the teleology and ideal types but with universal common descent, and of course with allowances for the many exceptions to the rule that generalized features precede the specialized in development. Then, too, the types have to be reinterpreted as ancestors, and the direction of development must go not from the generalized features of the class to the specialized features of the family, genus, and species, but rather from a generalized ancestral form, through increasingly specialized intermediate forms, to the present-day adult. Ironically, the more one Darwinizes Baer along these lines, the more he comes to resemble Haeckel.[66] In fact, Haeckel himself developed just such a Darwinized interpretation of Baer in a series of papers that modified or appropriated one part of Baer's system at a time.

Baer had not ruled out evolution entirely, but restricted its operation to minor changes within the boundaries of his four basic animal types. He also required that it be a kind of *Entwicklung*, a developmental process, teleological and law-abiding—not the result of blind and disorderly material processes like Darwinian variation and selection or Haeckel's external causes. However, Baer was also critical of linear developmental models and was able to find at least some common ground with the Darwinians. As he wrote to Haeckel in 1876, at the age of eighty-four:

My general views, which I have expressed here and there from 1834 on, were basically not very different from those of Darwin, but of course they were much more indefinite and uncertain. . . . Also, I never could call into doubt that the order of appearance of the various organisms [in the history of the Earth] must have been an *Entwicklung*. However, I could not fully agree with the formulation of Darwinism as it has appeared up till now, because for me, first and foremost, too much arbitrariness seems to dominate. Also, I could never bring myself to deny this *Entwicklung* a goal, without which the character of *Entwicklung* would be lost and the whole process would then merely be the result of a lot of petty physical factors [*Wirksamkeiten*].[67]

Clearly, Baer was trying to be conciliatory by saying his "general" views were similar to Darwin's. He probably was referring to their shared opinion that evolution did actually occur and that there was no linear progress from lower to higher. His other qualification, that evolution had to be an orderly and teleological *Entwicklung*, showed just how far his standpoint remained from Darwin's and especially from Haeckel's, who had been campaigning against *Entwicklung* in Baer's sense for the previous ten years. That campaign evidently had not made much of an impression on Baer.

Although he rejected developmental law and teleology, Haeckel was able to incorporate into his own system some individual tenets of Baer's. The first was Baer's description of ontogeny as a process of differentiation within a type, rather than a progression of increasingly complex forms. Haeckel reinterpreted this in evolutionary terms. First, he identified Baer's generalized vertebrate type with an actual ancestral vertebrate whose general features were preserved by heredity and recapitulated in the early embryo. Similarly, successive stages of differentiation recapitulated more and more recent and specialized ancestors.[68] After the ancestral vertebrate would come the ancestral bird, for example.

Second, Haeckel had to reconcile the two accounts of the direction of development, Baerian differentiation from general to specific, and progress from lower to higher or ancient to modern on an evolutionary lineage. Bronn had shown the way to make them commensurable using his analysis of morphological progress, which Haeckel repeated in *Generelle Morphologie*. Progress occurred by the following means: "1) Differentiation of functions and organs; 2) reduction in the number of organs of the same name; 3) concentration of functions and their organs in specific parts of the body; 4) centralization of . . . organ systems . . . ; 5) internalization, especially of the most refined organs, insofar as they do not necessarily need to occur on the surface in order to support the organism's relationships with the external world; 6) greater spatial extension."[69]

Of these, the main measure of progress, for Haeckel, as for Bronn, was physiological division of labor, or increasing specialization and differentiation of organs and functions. If specialization tended to confer Darwinian fitness, as Bronn (and, for that matter, Darwin) had already suggested, then, under natural selection, lineages would become more specialized and morphologically higher over time: "In the struggle for life, natural selection picks out those individuals for propagation that were best able to adapt themselves to the conditions of existence. Since, in most cases, those individuals are the best and most perfect, the process is necessarily connected, in general (excepting individual special cases!) with slow, but steadily effective *perfecting*, with *progress in organization*"[70] (emphasis in original). The direction of evolution, then, was simultaneously toward improved Darwinian fitness and morphological perfection, as measured by specialization, division of labor, and other criteria. There was therefore no conflict between differentiation and progress. The embryo, like the lineage, could be seen simultaneously as differentiating and as adding progressively more modern features. The causes of development were still different from Baer's, of course, but its directionality was more or less the same. The generalized features of the type or ancestor appeared early, and the specialized features of ever narrower subtypes or descendants developed subsequently.

If one accepted Haeckel's identification of the type with the ancestor and division of labor with both differentiation and progress, there was no dichotomy between Haeckel's and Baer's versions of recapitulation. The single remaining problem was that Baer had the embryo differentiating only within its type—the chick, for example, was a vertebrate from the start and never had to become one. But as we shall see below, Haeckel turned in the 1870s to a new project of unifying the Baerian (or Cuvierian) types and relating them to a universal animal ancestor and a corresponding universal embryonic stage, the gastrula.

Haeckel's faith in progress predated his Darwinism and did not derive from it. He considered progress an observable fact, which demanded a monistic, materialistic explanation, lest it mislead people into belief in teleology or divine providence:

In the past few decades the validity of the law of progressive development has been acknowledged from various sides as an empirically established fact, but the fact remained for most people a mysterious and incomprehensible "organic law of nature." Its explanation only seemed possible under the dualistic assumption of a teleological plan of Creation, which the Creator followed as he fabricated the organisms. A scientific, that is, a monistic, mechanistic-causal explanation of this empirical law first became possible through the theory of common descent and ultimately through that theory's basic causal idea, the selection theory.[71]

He stopped short of making progress a necessity or a necessary consequence of Darwinism, however. It was merely an "empirical law" or an observed trend, and the likely historical outcome of cumulative selection.

The conception of progress as differentiation also had the advantage of being applicable in both organic evolution and human culture. It became part of Haeckel's scheme for unifying the biological and the human sciences. In an 1868 lecture, he asserted that

Yes, even for the entire evolution of human cultural life, division of labor is of such fundamental importance that one could almost use the extent of the latter as a yardstick for the stage of development of the former. The wild, natural peoples who to this day are stuck at the lowest level lack division of labor along with culture, or else the division is limited to the differing occupations of the two sexes. In contrast, a main cause of the tremendous progress our cultural life has made in the past fifty years could almost be found in the extraordinarily high degree of our modern division of labor, namely in the realm of the sciences and their practical applications.[72]

But in human affairs, as in embryology and evolution, there were exceptions and trade-offs between specialization and other forms of progress or the demands of adaptation. Haeckel had reservations about making division of labor his only measure of human progress. The specialist could sometimes be overly narrow: "On the other hand, it is not to be forgotten that every division of labor necessarily brings, along with its quite predominant advantages and advances, also its great disadvantages and regressions in immediate consequence. We see this everywhere in the polymorphism of human society, which shows us in its political, and social, but especially in its scientific [wissenschaftlich—including natural science and humanities] development the most complicated and compound of all phenomena of differentiation."[73]

In the same context, Haeckel took a swipe at any recalcitrant colleagues who still rejected Darwinism, calling them prime examples of how specialization could be counterproductive. The narrow-minded specialists could no longer see the big picture: "We only have to cast a glance at the science of organismal morphology in its present sorry state in order to see the pronounced dark side of very advanced division of labor before our eyes. . . . If it were not so, then the theory of natural selection, the greatest step forward in human science this century, would have to dominate all of biology. The greatest disadvantages for science come about because most works are specialized on a very little field and adapted to the narrowest specialists' points of view, while no longer concerning themselves with the larger whole."[74]

The Gastraea Theory

Haeckel next turned against Baer's (and Cuvier's) system of the four fundamental animal types. The system, he argued, was already losing some of its grip on zoologists. In the 1840s the radiate type had been split into echinoderms and coelenterates, the articulates into arthropods and segmented worms, and the protozoa had been recognized as a separate type, making a total of seven instead of four. That indicated to Haeckel that the systematists could not maintain the Baerian distinctions, but were still clinging to the type concept and revising it instead of accepting a fully evolutionary interpretation.

Haeckel objected not so much to the rigidity of a system that would allow evolution only within a type, for he said he could accept four or seven or more separate evolutionary trees if the evidence favored such interpretations. Haeckel's main target was the assumption that a type corresponded to a transcendent plan or idea. He therefore built an elaborate case against "[the] immanent, original 'body plan [*Bauplan*] of the types,' which makes up the actual *fundamental teleological principle of the type theory*"[75] (emphasis in original).

Beginning with his monograph on the calcareous sponges, Haeckel repeatedly called attention to the ubiquity of certain early developmental stages in all the supposedly incommensurable types. First came the single cell of the fertilized egg, which divided itself up into the many-celled, mulberry-like "morula." Out of the morula, the hollow ball of cells called the "blastula" developed. Then, in Haeckel's ideal case, one side of the blastula became invaginated and a cuplike "gastrula" resulted. Haeckel focused his attention on the gastrula stage, which he claimed recapitulated an actual common ancestor of all the multicellular animal groups and which he named *Gastraea*. He touted the discovery of *Gastraea* as definitive evidence for the common descent of all metazoans, and against the separate Baerian types.[76]

The lowest animals, such as the sponges and coelenterates that were among Haeckel's specialties, got hardly more complex than the *Gastraea* level. The open end of the cup became the mouth, and a ring of tentacles might grow around it, as in the polyps and jellyfish; or the body cavity might become more convoluted, complex, and perforated, as in the sponges. Based on their common gastrula stage and the principle of recapitulation, Haeckel therefore argued that the sponges and the coelenterates were very closely related to each other and to *Gastraea*. He also identified some living species of sponges as being especially close to *Gastraea* in appearance.

Haeckel then proceeded to extend the argument from the unity of the sponges and coelenterates to the unity of all the major animal groups. There were some difficulties in the details, however, because the gastrula did not always form by invagination. In fact, even among species that he knew had to be closely related (e.g., among the sponges), there were different ways of producing a cup-shaped gastrula out of a spherical blastula. However, to Haeckel, the main thing was that they all had a blastula and a gastrula and some developmental pathway from one to the other. The varying mechanics of gastrulation were merely the result of adaptive changes that obscured part of the record of recapitulation.[77]

Haeckel presented the *Gastraea* theory as the continuation of his evolutionary reinterpretation of Baer. *Generelle Morphologie* had explained away the Baerian type in ontogeny as a heritage from a distant ancestor, the founder of each group. Now the *Gastraea* theory linked all those distant ancestors into one big phylogenetic tree, with a gastrula-like organism at its base. He wrote that the Baerian type had now been reduced to a consequence of heredity, and the grade of differentiation to a consequence of adaptive change. Thanks to the *Gastraea* theory, the types had been unified and the last remaining Baerian obstacle to recapitulation removed: "*Only through the Gastraea theory and its consequences is the phylogenetic relationship of the animal types to one another completely illuminated*"[78] (emphasis in original). There was no reason to distinguish between a type and any other taxonomic unit within a type. They were all simply various-sized portions of a genealogical tree.

Because of common descent from *Gastraea*, recapitulation no longer had to be restricted to within a single traditional type. The chick embryo, for example, could recapitulate its invertebrate ancestry, too, before reaching the generalized vertebrate stage. Recapitulation could go all the way back to *Gastraea*, if not further, to the simple colonial organisms and single cells from which *Gastraea* itself must have descended.

Haeckel's magnum opus of 1866 had set forth a comprehensive agenda for a historical approach to every aspect of biology, in which teleology and type would give way to open-ended, branching evolution, driven by unpredictable forces outside the organism. From 1866 through the 1870s Haeckel set about systematically demonstrating how evolutionary history and ancestral forms could be reconstructed from comparative studies of living forms and especially from embryological evidence. The *Gastraea* theory was a central component of his overall argument, not only for the common descent of all metazoans, but against Baerian typology and teleology, and for the practicability of a Darwinian science of life.

Conclusion

The historiography of German morphology and evolutionism has been plagued by a certain Whiggish condescension, especially toward H. G. Bronn and Ernst Haeckel, for supposedly clinging too long to transcendental concepts of type, developmental law, and progress, and resisting historically contingent variation and adaptation. Of course, the end point of the Whig trajectory has changed since the days when E. S. Russell could relegate all of Darwinism to the dustbin of history by associating it with the older morphological ideas. More recent versions tend to let Russell's association stand only in the case of German Darwinism, while having the Anglo-American variety triumph over the same conceptual obstacles and give rise to the modern evolutionary synthesis. Some correctives are available, but none that completely throws off Russell's influence. The literature consistently overstates the predominance of the old transcendentalism and its persistence in Germany through mid-century, and connects Bronn and Haeckel to it with little justification other than their language of law, development, type, and perfection.

Those biologists' explicit rejection and redefinition of transcendental archetypes and development as causes or explanations of historical change have seldom been taken very seriously, and neither have their concerns with the signature Darwinian themes of variation, adaptation, distribution, and historical contingency. In this study, close attention to the texts, the translation process, and especially the intellectual context in which the *Origin* was written and interpreted, yields a very different picture, not only of Bronn and Haeckel but of the older morphology in which they were supposedly mired.

Even during the so-called transcendental period, morphologists had hardly been unanimous on the meanings of type, the nature of developmental laws and forces, or any particular plan or scale of life. Beginning before the turn of the nineteenth century, German morphologists had been developing a wide range of interpretations. They did not all turn a blind eye to variation and diversity in order to focus on idealized types and rigid laws of development and recapitulation. On the contrary, they commonly wrote of *Mannigfaltigkeit*, or the diversity and variety of forms, as a phenomenon that needed to be appreciated and explained. Johann F. Meckel, for example, concerned himself extensively with the production of diversity, including differences between taxonomic groups, between the sexes, among normal individuals, and between normal and pathological cases. Other researchers, under the influence of Alexander von Humboldt, took increasing interest in biogeography and the roles of historical contingency and adaptation to local conditions in distributing forms.

German biologists, almost all of them striving to establish their specialties as university-based disciplines during an era of university modernization and reform, were much more conscious than their gentlemanly British counterparts of the need to uphold certain scholarly ideals and formalisms in their work. There was much less talk of divine plans and designs, and more rhetoric about upholding the standards of *Wissenschaft*, the proper systematization and illumination of biological knowledge, and the goal of deriving laws of organic nature. But the talk of law and order cannot be taken at face value. One could not hope to derive the myriad, *mannigfaltige* forms in nature from a small number of laws if the laws were kept rigidly deterministic, or if the patterns of development had to be perfectly linear and predictable. The developmental laws and forces of Johann F. Blumenbach, Carl F. Kielmeyer, and Johann C. Reil were designed to produce diverse results. There were always multiple laws of development, and they either interacted with the environment or worked at cross-purposes, some advancing and perfecting forms, others diversifying them or adapting them to changing circumstances, some enforcing conformity to type, and others exploring morphological possibilities within or between types. Progress, whether in the developing embryo, on an abstract scale of nature, or in the history of life, was not necessarily linear, and multiple criteria of progress were discussed. Types were not always transcendental, but could also be modeled after Cuvier's functional designs; or, as in the case of Meckel, they could be mere "collections" of individuals, grouped by general resemblances.

Baer's tongue-in-cheek "history" of recapitulationism and transformationism has been exposed to be just that. In 1828, he caricatured his predecessors in comparative embryology as taking a naive view of the embryonic stages of higher organisms as a sequence of adult forms from lower in the animal scale. He poked fun at species transformationists for uncritically taking over the same sequence and making evolution-ary change run along the same ridiculously inflexible track. Baer claimed credit for a more sensible view of development as differentiation, from a homogeneous or generalized early embryo to the more recognizable member of a class, order, family, genus, and species, with its increasingly specialized organs. He denied that an embryo or historical lineage could ever rise within a hierarchy of classes.

But this was not even intended to be an accurate description of either the recapitulationism or the transformationism of the time. The recapitu-lationists usually related embryonic development to scales of complexity or the hierarchy of taxonomic classes, and not at all to the history of spe-cies transformation. Even transformationists like Meckel did not make the connection. They did not derive their transformationism from their embryology, and they did not see evidence of common ancestry in the resemblances between embryonic forms. Moreover, Meckel's embryol-ogy was no more linear than Baer's. He had clearly worked out a descrip-tion of development as differentiation of initially diffuse and ill-defined structures.

By the time Bronn began developing his own approach to making *Wissenschaften* of zoology and paleontology in the 1840s and 1850s, German biology had a broad palette of methods and concepts to offer him. Like Darwin, he aimed for a historical conception of life, with a focus on adaptation and environment as determinants of form. Unlike Darwin, he embraced the need to uphold traditional ideals of *Wissenschaft* and accordingly to preserve the orderliness of the taxonomic system and to derive the interacting laws of form and change.

Bronn's starting point was already far removed from the idealism caricatured by Baer and Russell. In his principal pre-Darwinian works, he revived the problem of *Mannigfaltigkeit*, raised new concerns about variation, adaptation, the dynamics of interorganismal and organism-environment interactions, and offered concrete criteria by which to judge progress and perfection. He also used terms like "type," "developmen-tal law," and "perfection" in new ways. Eschewing transcendentalism, Bronn used Cuvierian functional types for the larger classes. Species, in contrast, were defined by descent from or resemblance to "prototypes," which were actual individuals, the founding members of the species. Various lawlike mechanisms saw to it that species would not vary too far, for too long, from their prototypes. When new species appeared in the fossil record, Bronn's law of unity of type made them resemble old ones, but higher laws of progress, diversification, and highest of all, adaptation, worked against type. The Earth's physical evolution necessitated continual adaptive improvement and progress of species, demanding diversity and change even as the law of type demanded continuity.

Because historical change was driven by external causes and the need to adapt to them, its course could not run parallel to that of embryonic development, according to Bronn. This was chiefly because the embryo had to become adapted, by stages, only to the present-day environment, not to a series of past ones. Embryonic stages might often look like lower or more ancient forms, but that was only because both the embryo and the fauna had to increase in complexity, division of labor, and so on, as they went from comparably simple origins to comparable present-day endpoints. But the specific changes observed in the embryo did not recapitulate any actual history. The past had to be understood through the study of fossils and geology, not through comparative embryology.

When Bronn wrote of law, perfection, and type, he clearly did not mean the same things as transcendental morphologists two or three decades earlier. The words should not be taken to mark him as hopelessly behind the times in the dawning Darwinian age, any more than Darwin's own discussions of the "law of Unity of Type," "progress," or "perfection" should similarly mark him. The corresponding words in the translation of the *Origin* cannot plausibly be read as twisting Darwin into conformity with outmoded views that Bronn himself had never even espoused. Rather, we should recognize that Bronn engaged with Darwin and explained him to German morphologists in their own language.

The result was much closer to Darwin's intentions than is ever admitted, but at the same time it changed Darwin's drift in subtle ways. Bronn accepted the concepts of struggle and selection, and he approved of Darwin's emphasis on the natural necessity for individuals and species to adapt continually to a changing environment. But he also toned down the imagery and metaphor that seemed to personify natural selection as an ersatz Designer, and he tried to accentuate the rule of law, but also the unpredictably complex interplay of laws in Darwin's system. Far from distorting Darwin beyond recognition, this did him a service by adapting his argument to German scholars' expectations for a *wissenschaftlich* presentation.

Still, Bronn was not just any German professor. He was a leading paleontologist with his own prizewinning theory of organic history to defend. Inexplicably, Darwin had failed to apply his vaunted networking skills in his dealings with Bronn. He displayed a surprising ignorance of Bronn's work and slighted Bronn's contributions to morphology and paleontology repeatedly, without even seeming to realize what he was saying. This elicited a few bilious responses from Bronn, and might have prompted Bronn to be more assertive of his own views in his commentary on the translation. He treated Darwin's ideas, where possible, as extensions of his own, particularly in connection with the primacy of environmental change and adaptation as causes of form. Bronn also sought common ground with Darwin in the analysis of progress and perfection, to which Bronn had devoted a great deal of effort. Darwin seemed to suggest himself that division of labor and specialization brought competitive advantages with them, and this gave Bronn the key to reconcile the two definitions of progress.

Bronn also questioned some of Darwin's particular arguments. Darwin's argument by analogy to artificial selection fell flat with Bronn, for some of the very reasons that had made it work with his British audience. These included the appeal to British authorities like John Herschel, who had called for any proposed new causes to be justified by analogy with known ones; Bronn cared not a whit for Herschel's methodological prescriptions. Darwin's analogy also banked on the familiarity with hunting dogs, fancy pigeons, and other examples on the part of his readers, and perhaps also on the parallel structure of William Paley's analogical argument about watches and organisms, which had been very successful. But here, again, the allusions were too Anglocentric to make much of an impression on the German professor.

Darwin's inferences about past events did not convince Bronn, either. They seemed to him to amount to special pleading, when laws of nature should have been invoked. Darwin had no trouble accounting for diversity of forms and infinite intergradations between them, but by the same token, he offered too little to preserve the integrity of species and higher taxonomic groups. Species stayed well-defined because natural selection kept eliminating hybrids and intermediates, but there was no guarantee, no law, that natural selection always had do so. Darwin could always tell some hypothetical evolutionary story about how the intermediates were eliminated, or else not, as the situation required; and this sort of tactic was not scientific enough for Bronn.

Finally, Bronn rejected Darwin's conclusion that successional theories like his own were obsolete. Admittedly, Bronn still had a mysterious creative force bringing new species into existence, and he was sometimes apologetic about this. But, as Bronn pointed out, Darwin needed such a force, too, at least for the very first species. Strictly speaking, natural selection had not fully supplanted it, and Bronn saw no reason why he could not continue to invoke it in his own theory. What difference did it make that he let it create more species, more often, than Darwin?

Even though he was ambivalent about Darwin's overall argument and felt that the case for the unlimited transformability of species had failed, Bronn played important roles in the development and reception of Darwin's theory. He was both a source on fossil biogeography, variation, adaptation, and gradual, progressive change, and a disseminator and interpreter of Darwin's views. His *Origin* was the first on the market in any language other than English. It was the version read by most German biologists, and probably by quite a few in other countries as well.

At the very least, an improved understanding of Bronn's translation helps explain some of Ernst Haeckel's interpretations and turns of phrase in his further development of Darwinism in the 1860s and 1870s, for Haeckel initially studied Bronn's *Origin*, not Darwin's. As he made his career as a promoter and popularizer of Darwinism, he was selling a product on which Bronn had left a distinct mark. He took over Bronn's language, and definitions, of progress, perfection, and law. He continued to connect division of labor to Darwinian competitive improvement. He also continued Bronn's work toward loosening the constraints of transcendental and other sorts of types on historical and geographic variation and change. He argued that the explanatory roles previously played by types could now be taken over by evolutionary common ancestors or common characteristics of a lineage. He used the terms more or less interchangeably, but not because he was trying to save the old concept of type. Darwin and Bronn had redefined "type" for him, and he used the word freely in its new sense.

But where Bronn had still been ambivalent about Darwin, Haeckel took action to eliminate the very problems Bronn had identified. He was most diligent in defending a naturalistic account of the appearance of the very first species, without a Creation or a creative force, as well as in further emphasizing the lawfulness of Darwinian processes. For the devotee of *Wissenschaft*, Haeckel provided laws in abundance, all working at cross-purposes, like Bronn's or even Meckel's, to maintain unity while also generating variety.

Of course, not everything in Haeckel derived from Bronn. He had goals, observations, and an ideology of his own. It is also significant that Haeckel was working on the systematics of the Radiolaria when he first encountered the *Origin*, rather than on comparative embryology. His initial interest was in Darwin's argument about the workings of the species "manufactory" and the distribution of varieties among species and of species among genera. Success should breed success, under natural selection, and Darwin had predicted that the greatest number of varieties, or incipient species, should be found in the genera that had already proven successful in producing species. Haeckel was pleased to be able to verify Darwin's prediction with his Radiolaria data. Thus the old problem of *Mannigfaltigkeit* was what first brought Haeckel around to Darwinism, not any perceived parallels between embryology and evolution. He worked those out subsequently, as a young professor at Jena, where he explored the opportunities for applying and formalizing Darwin's theory. He did not content himself with making modest reforms in morphology, but established an ambitious research program geared toward reconstructing phylogenetic relationships and unifying the life sciences and the human sciences within an evolutionary framework, while also promulgating the tenets of his monism.

Haeckel referred continually to his monism and anticlericalism, which were important considerations in every interpretive problem. He would brook no "dualism," no separate realm of types or ideas, mind or spirit. There was no room in his system for the will or goal of the organism to modify its form. He could tolerate no Creation or unique creative moment. Nature had to be continually creative, throughout evolutionary history, rather than merely working out preexisting potential. Hence Haeckel embraced external, environmental factors as the causes of variation and historical change. These broke up the chain of purely biological causes that might otherwise have stretched all the way back to the Creation. On such external and contingent causes of change, he found common ground with both Bronn and Darwin.

The breadth of Haeckel's program made special demands on the evolutionary process. There had to be strong, conservative laws and mechanisms of heredity to preserve the unity of type and remnants of old ontogenies, from which to reconstruct phylogenies. On the other hand, Haeckel needed to account for *Mannigfaltigkeit* and to have it emerge unpredictably over evolutionary time, rather than all at once at the Creation. Thus he also needed variation to break up hereditary unity and continuity, even if it meant falsifying the historical record preserved in the embryo. And the variations had to have mechanistic causes in the physical environment, not psychic, teleological, or providential ones. This was his reason for insisting on environmental effects and inheritance of acquired characteristics as the sources of adaptive variation. This version of the "Lamarckian" mechanism (i.e., without the perceived needs of the organism or any innate drive to perfection) was the only one available that was "mechanistic" in the proper sense for Haeckel's purposes. The appeal of the inheritance of acquired characteristics was not that it would make variation occur in adulthood and facilitate "terminal addition" of new features at the end of ontogeny. Haeckel used "Lamarckian heredity" for new, Darwinian purposes.

Darwinism, Old and New

This new picture of Haeckel has implications for our understanding of later developments in evolutionary thought, which I will outline only briefly here. It sheds new light on Haeckel's conflicts over "mechanistic" and experimental approaches to embryology, for example with Wilhelm His or Wilhelm Roux, and his rift with August Weismann over the latter's neo-Darwinism. It also makes Haeckel's lasting influence more apparent, because it makes Haeckel stand for other things besides inheritance of acquired characteristics and reactionary morphology. We can find evolutionists at the turn of the twentieth century or later who kept alive Haeckel's concerns to balance law and determinism with contingency and creativity.

One of the earliest and most sustained efforts to subvert Haeckel's program and upset his balance was made by the embryologist Wilhelm His, who offered a new way of looking at the mechanics of embryological development and the causes of morphological change in the embryo. He rejected Haeckel's historical explanations of those same morphological changes, along with the very idea that phylogeny could be the cause of ontogeny. His nasty dispute with Haeckel is usually seen as a battle over the value of experimental methodology, the status of embryology as a discipline and whether it was to be a mere handmaiden to Haeckelian phylogenizing. His is also counted as a forerunner of *Entwicklungsmechanik*, the turn-of-the-century study of the immediate or proximate causes and mechanisms of morphological change, which might include, for example, differential growth rates, movements of cells, chemical and physical signals.[1] But Haeckel's biggest problem with His was that he sought to eliminate historical explanations from biology. He reduced evolution, heredity, and adaptation to uninteresting by-products of the unchanging laws that made organisms grow, develop, and reproduce—all mechanical laws, such as those that governed the pushing and pulling of sheets of embryonic tissue into new shapes. In so doing, His threatened to upset Haeckel's balance of causes, and that is why Haeckel's response was so furious. To Haeckel, His represented an extreme of law and determinism that would reduce ontogeny and phylogeny to a mere running through of sequences that had been preordained at the Creation and were unresponsive to environmental change and the need for adaptation.

Challenging Haeckel's practical program in the 1870s, His declared the total independence of embryology and comparative anatomy from phylogeny. Any morphological resemblance between embryos or adults of different species proved only that they were products of the same or similar mechanical processes and initial conditions of development. It could not be taken as evidence of a hereditary relationship: "To reason from correspondence in type and developmental history to blood relationship should, from this moment on, no longer be allowed, since the prospect is opening up for understanding the various directions of development as exhaustive manifestations of a mathematically definable circle of possible ways of growing."[2] Haeckel's biogenetic law was a special object of His's scorn. No proper law of nature came with exceptions already built in, and it misled people into thinking they could explain ontogeny by inventing baseless histories.[3]

In response, Haeckel depicted His's system as a revised and expanded preformationism, in which not only the individual embryo but also the whole evolutionary lineage unfolded mechanically from a preformed germ. The whole scheme could hardly be expected to work unless divine providence foresaw the survival needs of the lineage and set things up to make the right adaptations appear when appropriate. Haeckel rejected this out of hand and demanded a causal role in ontogeny for phylogenetic history: "*Phylogeny is the mechanical cause of ontogeny*. With this statement, the principles of our monistic understanding of organic development are clearly characterized. . . . In the future, every naturalist will have to declare himself for or against this fundamental proposition, if he does not satisfy himself in the study of biogeny [i.e., ontogeny and phylogeny] with merely wondering at remarkable phenomena, but instead strives to go beyond that to an understanding of their meaning."[4]

Haeckel saw no room to compromise with any ahistorical view of life. There were no two ways about it, because the only alternatives to creative evolution implied teleology or divine providence:

With this statement an unbridgeable chasm is defined, which separates the older, teleological and dualistic morphology from the newer, mechanistic and monistic one. If the physiological functions of heredity and adaptation are shown to be the sole causes of organic form, then therewith, at the same time, every source of *teleology*, of dualistic and metaphysical points of view, will be removed from the field of biogeny. The sharp opposition between the two guiding principles is thus clearly defined. *Either a direct, causal connection between ontogeny and phylogeny exists or it does not exist.* Either ontogeny is a condensed excerpt of phylogeny or it is not. Between these two assumptions there is no third one! Either epigenesis and descent or preformation and Creation![5] (Emphasis in original)

In Haeckel's dichotomy, His had to be on the side of preformation and Creation, because he traced the causes of ontogeny and phylogeny back to initial conditions in the distant past.

Just a decade after the dispute with His, Haeckel had many of the same reasons to reject Weismann's ideas about heredity and development, though he was much more polite to Weismann. In retrospect Weismann's work is usually taken as a milestone on the road to modern Darwinism because of his rejection of the inheritance of acquired characteristics. However, it must be appreciated that the inheritance of acquired characteristics was not only a Lamarckian device, but also a source of Darwinian variation—for Haeckel the only legitimate, mechanistic source. From Haeckel's point of view, the burden was therefore on Weismann to provide a better account of variation, one that could still keep teleology and divine providence at bay. Weismann's solution was a far cry from the modern one, despite the "neo-Darwinian" label it has been given.

Like His, Weismann thought about the mechanics of how the various body parts, including the reproductive cells, developed out of the fertilized egg. He thought in terms of cell lineages[6] rather than larger structural elements like sheets and tubes, and although he made more provision for environmental effects, variation, and historical contingencies,[7] he also made some of the same inferences as His. He argued that the development of the germ cells could always be traced back, through a long sequence of cell divisions and movements, to the fertilized egg. The rest of the body, or "somatoplasm" (*Körperplasma*), was not involved in the process. It was derived from separate cell lineages that developed parallel to the germ line and did not contribute to it materially. Weismann concluded that there was no way for characteristics acquired during development or adulthood by the somatoplasm to be communicated to the germplasm and that therefore environmental effects of the sort Haeckel relied on could not be heritable.

But a further consequence of Weismann's system was that the germ line, following its own developmental laws, could, in principle, be traced back to some primeval period when all of its stock of internal "determinants" must have been formed, and with them, all of the animal world's potential for variation and evolution. Weismann was well aware of that consequence. He dated the fixation of the determinants to the time of the origin of multicellular organisms, since single-celled ones could not have had separate, sequestered germ lines. Their determinants were still exposed to environmental influences and could vary in the manner espoused by Haeckel. But from the earliest metazoan stage on, the germplasm could only rearrange and, in some limited ways, enhance existing determinants, not create anything fundamentally new: "I think to myself that the heritable individual variability of the lowest single-celled organisms arose from the direct influence of various sorts of external factors. Then, from the individual variability that is now given I derive that of the metazoans and metaphytes, namely in such a way that it would be immortalized, enhanced, and combined over and over in new ways by sexual reproduction, which in the meantime had become the general mode."[8] Even if the germ line did not extend all the way back to Creation, but only to the evolution of the first metazoans, that was hardly enough to redeem it in Haeckel's estimation.

In 1893, Haeckel responded in print to the germplasm theory, which Weismann had restated at length in *Das Keimplasma* (The germplasm) the year before. Once again, Haeckel expounded upon the importance of external causes of variation, and he admonished biologists to think about the philosophical and theological consequences of their theories. If the environment did not call forth heritable novelties, if all future variations and adaptations were foreseen and prepared in a primeval germplasm, then some form of teleology or divine providence was implied. That alone sufficed for Haeckel to place Weismann in the company of Nägeli, Kölliker, Baer, Agassiz, and even Moses as an unscientific purveyor of Creation mythology.[9] Weismann gradually gave up the idea that the determinants were unchangeable except through combination. In fact, the revision was pretty far along by 1892, before the above diatribe, but Weismann had not gone far enough to make a difference to Haeckel. He was still reining in nature's creativity too tightly.

Darwinism in the Twentieth Century

The rivalry between Haeckel and Weismann to define and defend Darwinism continued into the twentieth century, but the result was not the polarization of evolutionary theory into a neo-Darwinian and a Lamarckian or pseudo-Darwinian camp, based simply on the acceptance or rejection of the inheritance of acquired characteristics. In addition to strongly directional orthogenetic approaches, saltational theories and also many forms of Lamarckism were being upheld. Some Lamarckians emphasized, with E. S. Russell, the power of life to adapt itself to changing environments; some emphasized, with Theodor Eimer, the directionality and momentum of environmentally induced change; and some emphasized, with August Pauly, the role of mind in shaping organic matter. There were also compromise theories such as the American neo-Lamarckism of Edward D. Cope and others, who combined environmental effects and inner orthogenetic drives with natural selection. And there was also the biometric approach, which sought to quantify variation and the effects of selection without taking a firm position on the mechanism of heredity or the causes of variation.[10] In this increasingly diverse and contentious mix, Haeckel's views continued to be strongly represented, but by people who did not necessarily identify themselves as Haeckel's friends and allies.

I think of this group as the "old-school Darwinians," a liberal translation of "*Altdarwinisten*," which is what Ludwig Plate called them, in contradistinction to the Weismannian "*Neudarwinisten*," or neo-Darwinians.[11] To call them "Haeckelians" might overstate their devotion to the man and his larger monistic philosophy, and there does not seem to be any other term in English for Darwinians of the non-neo persuasion.[12] "Lamarckians" blurs too many distinctions, and Bowler's "pseudo-Darwinians" amounts to a Whiggish rejection of their position.

In stark contrast to Weismann, Haeckel rested on his laurels after the 1890s, at least in regard to theories of evolution and heredity. Mendelism looked to him like just another theory of discrete hereditary units, like Darwin's pangenesis or Weismann's germplasm, with nothing special to recommend it over the others and presenting no great need for his attention.[13] He also had no use for the new experimentalism, whether in genetics or in embryology. He thought it would not prove anything that was not already well known from comparative morphology or experience with animal breeding.[14] And, as he himself wrote to Thomas H. Huxley, he had become much more interested in promoting his monist philosophy and fighting the clerics than in continuing his biological research. He decided to "fill up the rest of my days with general studies, particularly of monistic philosophy. The fight against clericalism and the Medieval stupefaction of our so-called 'educated' elite [*Gebildeten*] (—theologians, jurists, philologists, etc.—) is becoming ever more important."[15]

It was left to younger Darwinians of the old school to pick up the gauntlet and try different ways of adapting Haeckel's Darwinism to twentieth-century conditions, especially the new experimentalism, the rising tide of genetics, and the expectation that biologists would solve social problems through eugenics and other applications of evolution and heredity. Haeckel's student Richard Semon, for example, broke his lance against Mendelian genetics and developed his *Mneme* theory of organic memory as an alternative that would allow the environment to produce heritable, new characteristics.[16] Plate, who took over Haeckel's chair at the university of Jena and became an influential journal editor and commentator on new research, favored a compromise with genetics. He allowed only a portion of an organism's heredity to be under the control of chromosomes and Mendelian laws. The rest, he assumed, was mediated by the cytoplasm and allowed environmental effects to be transmitted as the old-school required.[17]

The notorious Paul Kammerer took a third tack and appropriated the new methods of experimental morphology and genetics in order to defend an old-school conception of Darwinism that he had learned from his friend Semon and his reading of Haeckel. Kammerer aimed to demonstrate experimentally that the environment could indeed induce adaptive changes, and that the changes gradually would become fixed in heredity in the form of Mendelian genes on chromosomes. Especially after World War I and the deaths of Haeckel and Semon, Kammerer took it upon himself to modernize and popularize old-school Darwinism as an alternative to Weismannism, Mendelism, and eugenics. He felt that the war had made it impossible for any reasonable person to believe that universal struggle could possibly be the key to human progress or to the origin of anything new and useful in evolution. In order to make his Darwinism relevant in the postwar period, he played down the role of struggle and selection, stressed cooperation and the inheritance of environmental effects as the primary mechanisms of change, and outlined an alternative eugenics, based on environmentally induced modifications instead of selection.[18]

Instead of shunting them aside as pseudo-, pre-, anti-, or otherwise non-Darwinian in spirit, historians need to situate Haeckel and the younger Darwinians in his orbit within a broad landscape of early-twentieth-century theories. The aging Haeckel, the old school, the neo-Darwinians, and various other factions in Germany and abroad together *constituted* the Darwinism of the time, rather than "eclipsing" it, as in the standard story.[19] What was there for them to eclipse? No single theory can be identified as the one true Darwinism that was lurking in the shadows, waiting for the opportunity to emerge and become synthesized with Mendelism.

I would like to suggest that the modern evolutionary synthesis brought together many more points of view than has been generally recognized and was well worthy of being called a "synthesis," as opposed to a "constriction" that mainly ruled out competing ideas.[20] In particular, it is misleading to say simply that it ruled out "Lamarckism," if the term is meant to include old-school Darwinism. The ideas of the old school did not all stand and fall with the inheritance of acquired characteristics, and the synthesis developed positions in line with at least a few of those ideas. It rejected teleology, orthogenesis, and any role for mind and will in directing evolutionary change; it favored historical contingency and ascribed variation to "random" external causes, and it relied heavily on adaptation to explain progress and diversification. Also, even though it had little use for recapitulationism, it treated ontogeny as the product of evolutionary history rather than as playing any active role in initiating or directing evolutionary change. To claim a direct influence from Haeckel might be an exaggeration, because his reputation was at a low ebb at the time, but he did raise questions of lasting importance, which still demand answers today.

Translation and Transformation

Darwin's theory has appeared in many variations besides Bronn's and Haeckel's. Even Darwin's closest early supporters, such as Wallace, Huxley, Hooker, Lyell, or Gray, differed so much in their interpretations and applications of it that it has been impossible from the outset to identify a single, orthodox version.[21] In other national contexts the most prominent Darwinians read it first and foremost as a theory of social evolution and progress, inspiring movements toward national unity in Italy or against colonial domination in China.[22] Russian biologists grappled with the metaphor of struggle and shifted its range of meanings toward the inclusion of more cooperation and communal efforts to survive in harsh environments.[23] In its major twentieth-century incarnation, the modern synthesis, Darwin's basic mechanisms of variation and heredity were Mendelized and expressed in the mathematical language of population genetics, and evolution became identified with shifting gene frequencies. Later developments in molecular biology have kept the precise mechanisms and meanings of variation and heredity in flux, while proponents of "evo-devo" challenge the gene-centered interpretation altogether, harkening back to a time when morphology was the soul of natural history and evolution was about changes in form and function. Meanwhile, Darwin's "places" have morphed into ecological niches and the economy of nature into ecosystems. Plate tectonics has eliminated the steady Lyellian cycles of uplift and subsidence featured in Darwin's *Origin*. It has revolutionized biogeography and rendered some of Darwin's explanations of dispersal across geographic barriers superfluous. The inferred age of the Earth has grown tremendously. Mass extinctions by meteors, supervolcanoes, and other catastrophes beyond Darwin's worst nightmares have gained acceptance. Darwin's assumptions about progress and perfection in nature have become as obsolete as his overtures to Paleyan natural theology.

And yet, through all these changes in content and context, Darwinism somehow remains "Darwinism." Surely, then, the definition is flexible enough to include the versions of Bronn and Haeckel. They supplied interpretations of Darwin's model that kept it viable and made it applicable to problems deemed important in the German context of their day. We have seen Bronn drawing on earlier German sources and ideals of scholarship in order to fill out Darwin's account of progress and perfection, while also demanding a theory of ultimate origins, firmer rules of inference about historical events, and laws of change. We have seen Haeckel formulating new Darwinian rejoinders to Bronn's objections and demands, providing greater systematic organization of the facts of morphology, heredity, variation, and progress, and greater specificity about the origin of life and the course of phylogenetic history. Haeckel's determination to shore up the theory against teleology and directed evolution encouraged him to adopt Bronn's arguments for environmental causes of change, because they took the initiative away from the organism, its will, or any law of variation built into the lineage at the Creation. Haeckel balanced the historical contingency thus introduced with a concept of heredity that imposed historical constraints on development and preserved evidence of phylogeny in the living organism.

These interpretations, no matter how fleeting their successes might have been, are not to be set aside as uninteresting aberrations and intellectual dead-ends, or deplored as abuses of Darwinism. They are all part of the story of the development of Darwin's theory, a story not of steady progress toward today's understanding but of exploring the range of possible meanings and applications of Darwin's words and concepts. Darwin's gift to modern science, in the end, was not just "a theory by which to work," as he called it in his autobiography,[24] but rather a theory *on* which to work, and on which scientists still do work, in a continuing process of translation and transformation.

Notes

Introduction

1. For overviews of Darwin's life and work and his long delay, see, e.g., Francis Darwin, ed., *The Life and Letters of Charles Darwin: Including an Autobiographical Chapter*, 2 vols. (New York: D. Appleton, 1898), 1: 315–317, 347; Michael Ruse, *The Darwinian Revolution: Science Red in Tooth and Claw* (Chicago and London: University of Chicago Press, 1979), 184–185; Janet Browne, *Charles Darwin: Voyaging* (London: Jonathan Cape, 1995), 470–472; Adrian Desmond and James R. Moore, *Darwin: The Life of a Tormented Evolutionist* (1991; reprint, New York and London: W. W. Norton, 1994), 339–343.

2. A central theme for Janet Browne, *Charles Darwin: The Power of Place* (New York: Alfred A. Knopf, 2003); it also comes out well in Desmond and Moore, *Darwin*.

3. Letter of 15 October 1859 to T. H. Huxley, *Darwin Correspondence* 7: 350–351. See also the appendices on "Presentation copies of the *Origin*," *Darwin Correspondence* 7: 533–536 and 8: 554–570. Most were sent by the publisher, without a personal letter from Darwin.

4. Letter of 2 February 1860 to T. H. Huxley, *Darwin Correspondence* 8: 64–65.

5. Letter of 4 February 1860 to H. G. Bronn, *Darwin Correspondence* 8: 70.

6. Ibid.

7. On Darwin's exposure to these successional theories, see: Phillip R. Sloan, "Introductory essay: On the edge of evolution," in *The Hunterian Lectures in Comparative Anatomy, May–June, 1837* (Chicago: University of Chicago Press, 1992), 3–72; Phillip R. Sloan, "Darwin, vital matter, and the transformation of species," *Journal of the History of Biology* 19 (1986): 369–445.

8. Heinrich G. Bronn, *Handbuch einer Geschichte der Natur*, 3 vols. incl. atlas (Stuttgart: E. Schweizerbart, 1841–1849).

9. *Marginalia*, under "Bronn." Darwin appears to have begun reading it in 1845: Thomas Junker, "Heinrich Georg Bronn und die *Entstehung der Arten*," *Sudhoffs Archiv* 75 (1991): 180–208, on 187–188. Darwin's notes were mainly on

variation and artificial selection, but hybridization, dispersal, and other topics also attracted his attention. Bronn's successional theory, apparently, did not.

10. Heinrich G. Bronn, *Untersuchungen über die Entwickelungs-Gesetze der organischen Welt während der Bildungs-Zeit unserer Erd-Oberfläche* (Stuttgart: E. Schweizerbart, 1858).

11. Charles Darwin and Alfred R. Wallace, "On the tendency of species to form varieties; and on the perpetuation of varieties and species by natural means of selection," *Journal of the Linnean Society of London* 3 (1858): 45–62.

12. Letter of 18 [and 19] February 1860 to Charles Lyell, *Darwin Correspondence* 8: 92–95.

13. A tradition that follows the editorial comments by Darwin's son, Francis, in F. Darwin, ed., *Life and Letters of Charles Darwin*, 2: 70–71.

14. Walter Baron, "Zur Stellung von Heinrich Georg Bronn (1800–1862) in der Geschichte des Evolutionsgedankens," *Sudhoffs Archiv* 45 (1961): 97–109; Peter J. Bowler, *Fossils and Progress: Paleontology and the Idea of Progressive Evolution in the Nineteenth Century* (New York: Science History Publications, 1976); Junker, "Bronn und *Enstehung der Arten*"; Lynn K. Nyhart, *Biology Takes Form: Animal Morphology and the German Universities, 1800–1900* (Chicago and London: University of Chicago Press, 1995), 110–121; and Ingrid Schumacher, "Die Entwicklungstheorie des Heidelberger Paläontologen und Zoologen Heinrich Georg Bronn (1800–1862)" (Dissertation, Ruprecht-Karl-Universität, Heidelberg, 1975).

15. Letter of 2 February 1860 to T. H. Huxley, *Darwin Correspondence* 8: 64–65.

16. Found in the third (1861) and later editions of the *Origin: Variorum*, 59–70.

17. Peter J. Bowler, *The Non-Darwinian Revolution: Reinterpreting a Historical Myth* (Baltimore and London: Johns Hopkins University Press, 1988), 51, 82–90.

18. Paul Weindling, "Ernst Haeckel, *Darwinismus*, and the secularization of nature," in *History, Humanity, and Evolution: Essays for John C. Greene*, ed. James R. Moore (Cambridge: Cambridge University Press, 1989), 311–327.

19. Charles Darwin, *On the Origin of Species by Means of Natural Selection, or the Preservation of Favoured Races in the Struggle for Life*, 1st ed. (London: John Murray, 1859; facsimile, Cambridge, Mass. and London: Harvard University Press, 1964), 3.

20. Richard Dawkins, *The Blind Watchmaker: Why the Evidence of Evolution Reveals a Universe without Design* (1985; reprint, New York and London: W. W. Norton, 1987), 4–6.

21. Sander Gliboff, "Paley's design argument as an inference to the best explanation, or, Dawkins' dilemma," *Studies in the History and Philosophy of Biological and Biomedical Sciences* 31 (2000): 579–597.

22. Biographical information is from Schumacher, "Entwicklungstheorie des Heidelberger Paläontologen"; *ADB* and *DSB*, under "Bronn"; and *NDB*, 2: 633–634.

23. Bronn, *Geschichte der Natur*.

24. Scott L. Montgomery, *Science in Translation: Movements of Knowledge Through Cultures and Time* (Chicago: University of Chicago Press, 2000).

25. David L. Hull, "Exemplars and scientific change," *PSA—Proceedings of the Biennial Meeting of the Philosophy of Science Association* (1982), 2: 479–503; David L. Hull, "Darwinism as a historical entity," in *The Darwinian Heritage*, ed. David Kohn (Princeton: Princeton University Press, 1985), 773–812.

26. David N. Livingstone, *Putting Science in Its Place: Geographies of Scientific Knowledge* (Chicago: University of Chicago Press, 2003); Nicolaas A. Rupke, "Translation studies in the history of science: The example of *Vestiges*," *British Journal for the History of Science* 33 (2000): 209–222; Browne, *Charles Darwin: The Power of Place*.

27. Adrian Johns, *The Nature of the Book: Print and Knowledge in the Making* (Chicago: University of Chicago Press, 1998); James A. Secord, *Victorian Sensation: The Extraordinary Publication, Reception, and Secret Authorship of Vestiges of the Natural History of Creation* (Chicago: University of Chicago Press, 2000); James A. Secord, "Knowledge in transit," *Isis* 95 (2005): 654–672.

28. Paul Forman, "Weimar culture, causality, and quantum theory, 1918–1927: Adaptation by German physicists and mathematicians to a hostile intellectual environment," *Historical Studies in the Physical Sciences* 3 (1971): 1–115; Jonathan Harwood, *Styles of Scientific Thought: The German Genetics Community, 1900–1933* (Chicago: University of Chicago Press, 1993); Anne Harrington, *Reenchanted Science: Holism in German Culture from Wilhelm II to Hitler* (Princeton: Princeton University Press, 1996).

29. One might say, with Walter Baron, that Bronn was more like a historian and Darwin more like a sociologist of life: Baron, "Stellung von Heinrich Georg Bronn."

30. E. S. Russell, *Form and Function: A Contribution to the History of Animal Morphology* (London: John Murray, 1916), 89–91.

31. Ibid., 230–241, 247, on 247.

32. William Coleman, "Morphology between type concept and descent theory," *Journal of the History of Medicine and Allied Sciences* 31 (1976): 149–475; Mario Di Gregorio, *From Here to Eternity: Ernst Haeckel and Scientific Faith* (Göttingen: Vandenhoeck and Ruprecht, 2005), 191–197; Olaf Breidbach, "Darwinism and comparative anatomy in 19th century Germany," in *Giovanni Canestrini, Zoologist and Darwinist*, ed. Alessandro Minelli and Sandra Casellato (Venice: Istituto Veneto di Scienze, Lettere ed Arti, 2001), 149–167.

33. On Stöcker: Ann Taylor Allen, *Feminism and Motherhood in Germany, 1800–1914* (New Brunswick, N.J.: Rutgers University Press, 1991); Helene Stöcker, [Untitled], in *Was wir Ernst Haeckel verdanken: Ein Buch der Verehrung und Dankbarkeit*, 2 vols., ed. Heinrich Schmidt (Leipzig: Unesma, 1914), 2: 324–328. On Hirschfeld: Chandak Sengoopta, "Glandular politics: Experimental biology, clinical medicine, and homosexual emancipation in fin-de-siècle central Europe," *Isis* 89 (1998): 445–473; Magnus Hirschfeld, "Ernst Haeckel

und die Sexualwissenschaft," in *Was wir Ernst Haeckel verdanken*, 2: 282–284. On the socialists: Richard Weikart, *Socialist Darwinism: Evolution in German Socialist Thought from Marx to Bernstein* (San Francisco: International Scholars Publications, 1999). On the Freemasons: Fritz Bolle, "Monistische Maurerei," *Medizinhistorisches Journal* 16 (1981): 280–301. On the Nazi connection, see the discussion of Gasman, below.

34. Frederick Gregory, *Scientific Materialism in Nineteenth Century Germany* (Dordrecht: D. Reidel, 1977); Heinz Degen, "Vor hundert Jahren: Die Naturforscherversammlung zu Göttingen und der Materialismusstreit," *Naturwissenschaftliche Rundschau* 7 (1954): 271–277.

35. Monism is a constant theme in virtually all of his writings, from the 1866 *Generelle Morphologie* on, but see especially: Ernst Haeckel, *Die Welträthsel: Gemeinverständliche Studien über Monistische Philosophie* (Bonn: Emil Strauß, 1899); Ernst Haeckel, *Anthropogenie oder Entwickelungsgeschichte der Menschen*, 5th ed., 2 vols. (Leipzig: Wilhelm Engelmann, 1903); Ernst Haeckel, *Der Monismus als Band zwischen Religion und Wissenschaft*, 15th ed. (Leipzig: Alfred Kröner Verlag, 1911). For a recent overview, see Bernhard Kleeberg, *Theophysis: Ernst Haeckels Philosophie des Naturganzen* (Cologne: Böhlau, 2005).

36. That the cosmos as a whole might exhibit plan and purpose is not explicitly ruled out, and is sometimes strongly implied, but no additional plan or purpose distinguishes the organic world from the inorganic.

37. Heinrich Schmidt, ed., *Was wir Ernst Haeckel verdanken*.

38. Richard Goldschmidt, *Portraits from Memory: Recollections of a Zoologist* (Seattle: University of Washington Press, 1956), 31.

39. Erik Nordenskiöld, *The History of Biology: A Survey*, trans. Leonard Eyre (1928; reprint, New York: Tudor Publishing, 1936), 511–516.

40. Russell, *Form and Function*, 246–260, 302–308, 364.

41. Ibid., 257.

42. Exceptions are Nyhart, *Biology Takes Form*, 129–142; and to some extent, William M. Montgomery, "Germany," in *The Comparative Reception of Darwinism*, ed. Thomas F. Glick (Austin and London: University of Texas Press, 1974), 81–116.

43. Nordenskiöld, *History of Biology*, 518.

44. Daniel Gasman, *The Scientific Origins of National Socialism: Social Darwinism in Ernst Haeckel and the Monist League* (New York: American Elsevier, 1971); Daniel Gasman, *Haeckel's Monism and the Birth of Fascist Ideology* (New York: Peter Lang, 1997).

45. Stephen J. Gould, *Ontogeny and Phylogeny* (Cambridge, Mass.: Harvard University Press/Belknap, 1977), 77–78.

46. Bowler, *Non-Darwinian Revolution*, 51, 82–90.

47. Geoff Eley, ed., *Society, Culture, and the State in Germany, 1870–1930* (Ann Arbor: University of Michigan Press, 1996); David Blackbourn and Geoff Eley, *The Peculiarities of German History: Bourgeois Society and Politics in Nineteenth-*

Century Germany (Oxford: Oxford University Press, 1984). On Haeckel's liberalism and nationalism, see Ted Benton, "Social Darwinism and Socialist Darwinism in Germany: 1860–1900," *Rivista di Filosofia* 23 (1982): 79–121.

48. Timothy Lenoir, *The Strategy of Life: Teleology and Mechanics in Nineteenth Century German Biology* (Dordrecht: D. Reidel, 1982).

49. Nyhart, *Biology Takes Form*, 6–9.

50. Ibid., 66–102.

51. Robert J. Richards, *The Meaning of Evolution: The Morphological Construction and Ideological Reconstruction of Darwin's Theory* (Chicago: University of Chicago Press, 1992), 47–48, 167–169.

52. Ibid., 167–180, 90–143; Robert J. Richards, "Ideology and the history of science," *Biology and Philosophy* 8 (1993): 103–108; Robert J. Richards, *The Romantic Conception of Life: Science and Philosophy in the Age of Goethe* (Chicago and London: University of Chicago Press, 2002), 10, 514–554.

53. Di Gregorio, *Here to Eternity*, 105.

54. Ibid., 18–20, 191–197.

55. Richard Weikart, *From Darwin to Hitler: Evolutionary Ethics, Eugenics, and Racism in Germany* (New York: Palgrave Macmillan, 2004); but see Sander Gliboff, "Darwin on trial again," review of *From Darwin to Hitler: Evolutionary Ethics, Eugenics, and Racism in Germany*, by Richard Weikart (2004) (H-German, 2004), available from http://www.h-net.org/reviews/; and Richard Weikart, "Re: REV: Gliboff on Weikart, *Darwin to Hitler* (Weikart)," (H-German, 2004), available from http://h-net.msu.edu/cgi-bin/logbrowse.pl, in the H-German list, September 2004.

1 The Science of Life at the Turn of the Nineteenth Century

1. Lorenz Oken and Dietrich Georg von Kieser, eds., *Beiträge zur vergleichenden Zoologie, Anatomie und Physiologie*, 2 vols. (Bamberg: Joseph Anton Göbhardt, 1806–1807), 1: iii–iv.

2. Johann F. Meckel, "Entwurf einer Darstellung der zwischen dem Embryonalzustande der höhern Thiere und dem permanenten der niedern Statt findenden Parallele," in *Beyträge zur vergleichenden Anatomie* (Leipzig: Carl Heinrich Reclam, 1811), 2 (Book 1): 1–60, on 2.

3. Johann F. Meckel, *System der vergleichenden Anatomie*, 5 vols. (Halle: Rengersche Buchhandlung, 1821), 1: x.

4. From Jean-Baptiste de Lamarck, *Hydrogéologie, ou, Recherches sur l'influence qu'ont les eaux sur la surface du globe terrestre* (Paris: Chez l'auteur, 1802), as quoted and translated in William Coleman, *Biology in the Nineteenth Century: Problems of Form, Function, and Transformation* (1971; reprint, Cambridge: Cambridge University Press, 1977), 2.

5. Philip F. Rehbock, *The Philosophical Naturalists: Themes in Nineteenth-Century British Biology* (Madison: University of Wisconsin Press, 1983).

6. Coleman, *Biology in the Nineteenth Century*, 1–3; Nyhart, *Biology Takes Form*, 1; Lenoir, *Strategy of Life*, 1–2. The first usage of "biology" in this sense has been attributed to Treviranus (in 1802, in *Biologie, oder Philosophie der lebenden Natur für Naturforscher und Aerzte*, 6 vols., Göttingen: Johann Friedrich Röwer, 1802–1822), and Lamarck (in 1802, in *Hydrogéologie*); but priority apparently goes to Karl F. Burdach (in *Propädeutik zum Studium der gesammten Heilkunst*, Lepizig, 1800): Günther Schmid, "Über die Herkunft der Ausdrücke Morphologie und Biologie," *Nova Acta Leopoldina* NF 2 (1935): 597–620. On even earlier usages of the word, see Peter McLaughlin, "Naming biology," *Journal of the History of Biology* 35 (2002): 1–4. But see also Kanz, who argues that these early usages represented a false start for modern biology: Kai Torsten Kanz, "'. . . die Biologie als die Krone oder der höchste Strebepunct aller Wissenschaften': Zur Rezeption des Biologiebegriffs in der romantischen Naturforschung (Lorenz Oken, Ernst Bartels, Carl Gustav Carus)," *NTM—Internationale Zeitschrift für Geschichte und Ethik der Naturwissenschaften, Technik und Medizin* 15 (2006): 77–92.

7. On the disciplinary geography of the life sciences at the German universities, see Nyhart, *Biology Takes Form*.

8. Named after Wilhelm von Humboldt, brother of the naturalist Alexander. On Humboldtian educational ideals and their role in the founding of the Berlin University in 1910, see: Elinor S. Shaffer, "Romantic philosophy and the organization of the disciplines: The founding of the Humboldt University of Berlin," in *Romanticism and the Sciences*, ed. Andrew Cunningham and Nicholas Jardine (Cambridge: Cambridge University Press, 1990), 38–54; William H. Bruford, *The German Tradition of Self-Cultivation: "Bildung" from Humboldt to Thomas Mann* (Cambridge: Cambridge University Press, 1975).

9. On the German university and its research ideals, see: R. Steven Turner, "The growth of professorial research in Prussia, 1818 to 1848—causes and context," *Historical Studies in the Physical Sciences* 3 (1971): 137–82; R. Steven Turner, "The Prussian Universities and the Research Imperative" (dissertation, Princeton University, 1973); Friedrich Paulsen, *Die deutschen Universitäten und das Universitätsstudium* (Berlin: A. Ascher, 1902); Joseph Ben-David, *The Scientist's Role in Society: A Comparative Study* (Englewood Cliffs, N.J.: Prentice-Hall, 1971); and more recently, Thomas Broman, "University reform in medical thought at the end of the eighteenth century," *Osiris* 2nd series, 5 (1989): 36–53; Frederick Gregory, "Kant, Schelling, and the administration of science in the Romantic era," *Osiris* 2nd series, 5 (1989): 17–35. For a broader view that questions the university's monopoly on research as well as the actual independence of *Wissenschaft* from political and industrial money and interests, see: Peter Borscheid, *Naturwissenschaft, Staat und Industrie in Baden (1848–1914)* (Stuttgart: Ernst Klett Verlag, 1976); Kathryn M. Olesko, "Introduction [to volume on 'Science in Germany']," *Osiris* 2nd series, 5 (1989): 7–14; Arleen Tuchman, *Science, Medicine, and the State in Germany: The Case of Baden, 1815–1871* (New York and Oxford: Oxford University Press, 1994); Arleen Tuchman, "Institutions and disciplines: Recent work in the history of German science," *Journal of Modern History* 69 (1997): 298–319; Lynn K. Nyhart, "Civic and economic zoology in

nineteenth-century Germany: The 'living communities' of Karl Möbius," *Isis* 89 (1999): 605–630.

10. Thomas S. Hall, "On biological analogs of Newtonian paradigms," *Philosophy of Science* 35 (1968): 6–27; Lenoir, *Strategy of Life*, 18–22, 37–53.

11. Gliboff, "Paley's design argument."

12. Timothy Lenoir, "Kant, Blumenbach, and vital materialism in German biology," *Isis* 71 (1980): 77–108; Timothy Lenoir, "The Göttingen school and the development of transcendental *Naturphilosophie* in the romantic era," *Studies in History of Biology* 5 (1981): 111–205; Lenoir, *Strategy of Life*.

13. Johann F. Blumenbach, *Über den Bildungstrieb und das Zeugungsgeschäfte* (Göttingen: Johann Christian Dieterich, 1781; facsimile, Stuttgart: Gustav Fischer, 1971); Carl F. Kielmeyer, *Ueber die Verhältniße der organischen Kräfte unter einander in der Reihe der verschiedenen Organisationen, die Gesetze und Folgen dieser Verhältniße* (Stuttgart, 1793; facsimile, Marburg a. d. Lahn: Basilisken-Presse, 1993); Johann C. Reil, "Von der Lebenskraft," in *Gesammelte kleine physiologische Schriften* (Vienna: Aloys Doll, 1811), 1: 3–133 (reprinted from *Archiv für Physiologie* 1 [1796]: 8–162).

14. On the preformation-epigenesis debates, see: Shirley A. Roe, *Matter, Life, and Generation: Eighteenth-century Embryology and the Haller-Wolff Debate* (Cambridge and New York: Cambridge University Press, 1981); James L. Larson, *Interpreting Nature: The Science of Living Form from Linnaeus to Kant* (Baltimore: Johns Hopkins University Press, 1994).

15. Blumenbach, *Bildungstrieb*.

16. Kai Torsten Kanz, ed., *Philosophie des Organischen in der Goethezeit. Studien zu Werk und Wirkung des Naturforschers Carl Friedrich Kielmeyer (1765–1844)* (Stuttgart: Franz Steiner Verlag, 1994); Kai Torsten Kanz, "Carl Friedrich Kielmeyer (1765–1844)—Leben, Werk, Wirkung: Perspektiven der Forschung und Edition," in the same volume, 13–32; Kielmeyer, *Verhältniße der organischen Kräfte*.

17. Kielmeyer did not divvy up the life processes in quite the same way as Blumenbach, so the *Reproductionskraft* might not have had all of the same functions as the *Bildungstrieb*. At the very least it initiated the formative process, even if other forces had to make contributions during the course of development.

18. Kielmeyer, *Verhältniße der organischen Kräfte*, 9–12.

19. Ibid., 30.

20. Reil, "Lebenskraft," especially 36–45.

21. Susan F. Cannon, "Humboldtian science," in *Science in Culture: The Early Victorian Period* (New York: Science History Publications, 1978), 73–110; Malcolm Nicolson, "Humboldtian plant geography after Humboldt: The link to ecology," *British Journal for the History of Science* 29 (1996): 289–310.

22. Nicolaas A. Rupke, "Humboldtian medicine," *Medical History* 40 (1996): 292–310; Franz Weiling, "J. G. Mendel sowie die von M. Pettenkofer angeregten Untersuchungen des Zusammenhanges von Cholera- und Typhus-Massenerkrankungen mit dem Grundwasserstand," *Sudhoffs Archiv* 59, (1975): 1–19.

23. Janet Browne, *The Secular Ark: Studies in the History of Biogeography* (New Haven and London: Yale University Press, 1983); Franz Unger, *Versuch einer Geschichte der Pflanzenwelt* (Vienna: Wilhelm Braumüller, 1852).

24. Immanuel Kant, *Kants gesammelte Schriften,* Akademie edition (Berlin: Georg Reimer, 1902–1923; online version), 5: 400, available from http://www.ikp.uni-bonn.de/kant/.

25. Lenoir, *Strategy of Life.* On Kant's prescriptions for biology, see also Peter McLaughlin, *Kant's Critique of Teleology in Biological Explanation: Antinomy and Teleology* (Lewiston: E. Mellen Press, 1990); Richards, *Romantic Conception of Life,* 8–14.

26. Lenoir, "Göttingen school"; Lenoir, *Strategy of Life.* For more on Blumenbach's debts to Kant, see: Peter McLaughlin, "Blumenbach und der Bildungstrieb: Zum Verhältnis von epigenetischer Embryologie und typologischem Artbegriff," *Medizinhistorisches Journal* 17 (1982): 357–372.

27. William Coleman, "Limits of the recapitulation theory: Carl Friedrich Kielmeyer's critique of the presumed parallelism of earth history, ontogeny, and the present order of organisms," *Isis* 64 (1973): 341–350.

28. Richards, *Romantic Conception of Life,* 238–251.

29. Robert J. Richards, "Kant and Blumenbach on the *Bildungstrieb*: A historical misunderstanding," *Studies in the History and Philosophy of Biological and Biomedical Sciences* 31 (2000): 11–32; Richards, *Romantic Conception of Life.* Against Lenoir's interpretation, see also Kenneth L. Caneva, "Teleology with regrets," review of Timothy Lenoir, *The Strategy of Life, Annals of Science* 47 (1990): 291–300.

30. Johann F. Blumenbach, *Über den Bildungstrieb und das Zeugungsgeschäfte,* 2nd ed. (Göttingen: Johann Christian Dieterich, 1791), 32–34; Johann F. Blumenbach, *Handbuch der Naturgeschichte,* 11th ed. (Göttingen: Dieterich'sche Buchhandlung, 1825), 14–17.

31. Kielmeyer, *Verhältniße der organischen Kräfte,* 3.

32. Ibid., 4.

33. Ibid., 5.

34. Ibid., 5–6.

35. Richards interprets this same passage as a rejection of Kant's heuristic, as-if teleology and an endorsement of a genuine, constitutive purposiveness, but I cannot follow his reasoning: cf. Richards, *Romantic Conception of Life,* 241–242.

36. Darwin, *Origin of Species,* 434.

37. Ernst Mayr, *The Growth of Biological Thought: Diversity, Evolution, and Inheritance* (Cambridge, Mass.: Harvard University Press/Belknap, 1982), 45–47; Mary P. Winsor, "Non-essentialist methods in pre-Darwinian taxonomy," *Biology and Philosophy* 18 (2003): 387–400; Ronald Amundson, *The Changing Role of the Embryo in Evolutionary Thought: Roots of Evo-Devo* (Cambridge and New York: Cambridge University Press, 2005), 204–209.

38. Paul L. Farber, "The type concept in zoology in the first half of the nineteenth century," *Journal of the History of Biology* 9 (1976): 93–119; Gordon R. McOuat, "Species, rules, and meaning: The politics of language and the ends of definitions in 19th century natural history," *Studies in the History and Philosophy of Science* 27 (1996): 473–519.

39. Johann W. von Goethe, *Die Metamorphose der Pflanzen* (Gotha: Carl Wilhelm Ettinger, 1790; facsimile, Weinheim: Acta Humaniora, 1984).

40. Lorenz Oken, *Lehrbuch der Naturphilosophie*, 2nd ed. (Jena: Friedrich Frommann, 1831), 387–388; Thomas Bach, " 'Was ist das Thierreich anders als der anatomirte Mensch . . . ?': Oken in Göttingen (1805–1807)," in *Lorenz Oken (1779–1851): Ein politischer Naturphilosoph*, ed. Olaf Breidbach, Hans-Joachim Fliedner and Klaus Ries (Weimar: Verlag Hermann Böhlaus Nachfolger, 2001), 73–91.

41. Nicolaas A. Rupke, "Richard Owen's vertebrate archetype," *Isis* 84 (1993): 231–251.

42. Max Pfannenstiel, *Lorenz Oken, sein Leben und Wirken* (Freiburg im Breisgau: Hans Ferdinand Schulz Verlag, 1953); Oken, *Lehrbuch der Naturphilosophie*, 147.

43. Christian Pander, "Allgemeine Bemerkungen über die äußeren Einflüsse auf die organische Entwicklung der Thiere," in *Die Vergleichende Osteologie*, ed. Christian Pander and Eduard d'Alton, Sr. (Bonn: Eduard Weber, 1824), as reproduced in Boris Raikov, *Christian Heinrich Pander. Ein bedeutender Biologe und Evolutionist/An Important Biologist and Evolutionist, 1794–1865*, trans. Woldemar von Hertzenberg and Peter von Bitter (Frankfurt am Main: Waldemar Kramer, 1984), 35–43, on 39.

44. William Coleman, *Georges Cuvier, Zoologist: A Study in the History of Evolution Theory* (Cambridge, Mass.: Harvard University Press, 1964); Georges Cuvier, *The Animal Kingdom, Arranged After Its Organization*, trans. Edward Blyth et al. (London: William S. Orr, 1849).

45. Arthur O. Lovejoy, *The Great Chain of Being: A Study in the History of an Idea* (Cambridge, Mass.: Harvard University Press, 1936); Georg Uschmann, *Der morphologische Vervollkommnungsbegriff bei Goethe und seine problemgeschichtlichen Zusammenhänge* (Jena: Gustav Fischer, 1939).

46. Kielmeyer, *Verhältniße der organischen Kräfte*, 39.

47. For more on Kielmeyer's view of organic history, see Lenoir, *Strategy of Life*, 39–44; Kanz, "Carl Friedrich Kielmeyer."

48. On this point I support Kanz against Richards, who ascribes recapitulation of morphological features to Kielmeyer, the better to link him to Haeckel and Darwin. Richards reasons that forces and faculties can be expressed only in morphological structures, so that even though Kielmeyer writes exclusively about the former, he must mean the latter, too: Richards, *Romantic Conception of Life*, 243–246, 245n. But this does not follow, logically or biologically. In Kielmeyer's usage, a faculty such as "sensibility" might find expression in any number of different nervous systems, sense organs, or structures—eyespots, compound eyes,

ears, noses, brains. These would all count as parallel additions of the *faculty*, but without the morphological resemblance.

49. Friedrich Tiedemann, *Zoologie* (1808), quoted and translated in Richards, *Meaning of Evolution*, 43–44.

50. Gottfried R. Treviranus, *Biologie, oder Philosophie der lebenden Natur*, 6 vols. (Göttingen: Johann Friedrich Röwer, 1802–1822), 3: 4–5. On the world-organism see also 1: 34–50, 3: 40, 224–226.

51. Oken, *Lehrbuch der Naturphilosophie*, 387.

52. Rudolf Beneke, *Johann Friedrich Meckel der Jüngere* (Halle: Max Niemeyer, 1934).

53. Johann F. Meckel, "Ueber den Charakter der allmählichen Vervollkommnung der Organisation, oder den Unterschied zwischen den höhern und niedern Bildungen," in *Beyträge zur vergleichenden Anatomie* (Leipzig: Carl Heinrich Reclam, 1811), 2 (Book 1): 61–123, on 67; Meckel, *System*, 1: 10.

54. Meckel, *System*, 1: 5–6.

55. Ibid., 1: 10–11.

56. Meckel, "Charakter der allmählichen Vervollkommnung," 64; Meckel, *System*, 1: 228.

57. Meckel, "Darstellung der Parallele," 29–37.

58. Meckel, *System*, 1: 9.

59. Johann F. Meckel, "Beyträge zur Geschichte des menschlichen Fötus," in *Beyträge zur vergleichenden Anatomie* (Leipzig: Carl Heinrich Reclam, 1808), 1 (Book 1): 57–124, on 102–105; Meckel, "Darstellung der Parallele," 8–17.

60. Meckel, "Geschichte des menschlichen Fötus," especially 57–62, 79–95.

61. Meckel, *System*, 1: 7.

62. E.g., Meckel, "Geschichte des menschlichen Fötus," on 57–63; and especially Meckel, "Charakter der allmählichen Vervollkommnung," on 64.

63. Russell, *Form and Function*, 93.

64. Johann F. Meckel, "Besprechung dreyer kopfloser Missgeburten, nebst einigen allgemeinen Bemerkungen über diese Art von Missbildung," in *Beyträge zur vergleichenden Anatomie* (Leipzig: Carl Heinrich Reclam, 1809), 1 (Book 2): 136–162, on 158–159.

65. Meckel, *System*, 1: 396.

66. Ibid., 1: 416. Russell construed this passage not as Meckel's actual thesis, but as a tactical retreat from the unrealistic contention that Russell attributed to him, i.e., that the embryo as a whole recapitulated a lower adult: Russell, *Form and Function*, 93.

67. Meckel, *System*, 1: 222–227.

68. Ibid., 1: 341–350.

69. Karl Ernst von Baer, *Über Entwickelungsgeschichte der Thiere: Beobachtung und Reflexion*, 2 vols. (Königsberg: Gebrüder Bornträger, 1828–1837), 1: 147.

70. Ibid., 148.

71. Ibid., 1: 219.

72. Ibid., 1: 201n.

73. Richards, *Meaning of Evolution*, 47.

74. Meckel did note the difficulty of making comparisons between Cuvierian classes and said that his parallels were most distinct within classes: Meckel, "Darstellung der Parallele," 4–5.

75. Baer, *Entwickelungsgeschichte der Thiere*, 1: 204–205.

76. Meckel, "Darstellung der Parallele," 6–7, 24–6.

77. Baer, *Entwickelungsgeschichte der Thiere*, 1: 203–204.

78. In the more detailed survey by William Montgomery, not one of the German recapitulationists drew historical, morphological parallels: William M. Montgomery, "Evolution and Darwinism in German Biology, 1800–1883" (dissertation, University of Texas at Austin, 1974); Montgomery, "Germany."

79. Robert Chambers, *Vestiges of the Natural History of Creation*, 1st ed. (1844), as reprinted in *Vestiges of the Natural History of Creation and Other Evolutionary Writings*, ed. James A. Secord (Chicago: University of Chicago Press, 1994), 197–205, 212–213.

80. Robert Chambers, *Natürliche Geschichte der Schöpfung des Weltalls, der Erde und der auf ihr befindlichen Organismen*, trans. Carl Vogt, 2nd ed. (Braunschweig: Friedrich Vieweg, 1858).

81. Mary P. Winsor, *Starfish, Jellyfish, and the Order of Life* (New Haven and London: Yale University Press, 1976), 108.

82. Sander Gliboff, "Evolution, revolution, and reform in Vienna: Franz Unger's ideas on descent and their post-1848 reception," *Journal of the History of Biology* 31 (1998): 179–209; Franz Unger, *Botanische Briefe* (Vienna: Carl Gerold, 1852).

2   H. G. Bronn and the History of Nature

1. Matthias Schleiden, *Die Pflanze und ihr Leben: Populäre Vorträge*, 1st ed. (Leipzig: Wilhelm Engelmann, 1848), 1. On continuing concerns about the *wissenschaftlich* status of the life sciences throughout the 1840s or beyond, see Nyhart, *Biology Takes Form*, 63–102.

2. Matthias Schleiden, "Methodologische Einleitung [zur 4. Auflage der *Grundzüge der Wissenschaftlichen Botanik* (1861)]," in *Wissenschaftsphilosophische Schriften*, ed. Ulrich Charpa (Cologne: Jürgen Dinter, Verlag für Philosophie, 1989), 47–196, on 47; Anne Mylott, "The roots of cell theory in sap, spores, and Schleiden" (dissertation, Indiana University, 2002), 254–260.

3. Unger, *Botanische Briefe*, 3–6. On Unger's approach and its influence, see also Gliboff, "Evolution, revolution"; Sander Gliboff, "Gregor Mendel and the laws of evolution," *History of Science* 37 (1999): 217–235.

4. Nyhart, *Biology Takes Form*, 91.

5. Schumacher, "Entwicklungstheorie des Heidelberger Paläontologen," 25–26.

6. Bronn, *Geschichte der Natur*, 1: vi.

7. Ibid., 1: 3–5, 8–9.

8. Ibid., 1: 63–67.

9. Ibid., 1: vi.

10. Ibid., 3: 762.

11. Ibid., 3: 765–766.

12. Ibid., 3: 809.

13. Ibid., 3: 973; see also 3: 853ff.

14. Ibid., 3: 854.

15. Ibid., 3: 748.

16. Ibid., 3: 816.

17. Ibid., 3: 838.

18. Ibid.

19. Ibid., 3: 838–839.

20. Ibid.

21. On Darwin's comparable compromise, see Cannon, "The bases of Darwin's achievement: A reevaluation," *Victorian Studies* 5 (1961): 109–134.

22. Bronn, *Geschichte der Natur*, 1: 136ff.

23. Ibid., 1: 446.

24. Ibid., 3: 810.

25. Ibid., 3: 813–814.

26. Ibid., 3: 814.

27. Bronn, *Entwickelungs-Gesetze*, 112.

28. Bronn, *Geschichte der Natur*, 2: 9.

29. Ibid., 3: 973.

30. Ibid., 3: 840, 854; also 3: 974–975.

31. Ibid., 2: 62–64, 3: 747–748; quotes on 2: 63, 3: 747.

32. Despite its utter impracticability for the paleontologist, he later made interfertility his primary test criterion: Bronn, *Entwickelungs-Gesetze*, 232.

33. Bronn, *Geschichte der Natur*, 2: 85–180.

34. Desmond and Moore, *Darwin*, 247. See also James A. Secord, "Nature's fancy: Charles Darwin and the breeding of pigeons," *Isis* 72 (1981): 162–186; James A. Secord, "Darwin and the breeders: A social history," in *The Darwinian Heritage*, ed. David Kohn (Princeton: Princeton University Press, 1985), 519–542.

35. Bronn, *Geschichte der Natur*, 2: 180.

36. Ibid.

37. Ibid., 2: 181.

38. Bronn's precise claim is not documented; Darwin only mentioned it incredulously to Huxley: Letter of 2 February 1860 to T. H. Huxley, *Darwin Correspondence*, 8: 64–65.

39. Bronn, *Geschichte der Natur*, 2: 23–28.

40. Ibid., 2: 28.

41. Ibid.

42. Ibid., 2: 193–195. See also Bronn, *Entwickelungs-Gesetze*, 79.

43. Bronn, *Geschichte der Natur*, 2: 29.

44. Ibid., 3: 746.

45. Ibid., 2: 8.

46. Ibid., 2: 8, 3: 746–747.

47. For details of the competition: Schumacher, "Entwicklungstheorie des Heidelberger Paläontologen," 32–51; Bronn, *Entwickelungs-Gesetze*, III–V.

48. Bronn, *Entwickelungs-Gesetze*, 92–93.

49. Ibid., 93–96.

50. Ibid., 86.

51. Ibid., 96–104.

52. Heinrich G. Bronn, *Morphologische Studien über die Gestaltungs-Gesetze der Naturkörper überhaupt und der organischen insbesondere* (Leipzig and Heidelberg: C. F. Winter, 1858).

53. Ibid., 81, 83–108.

54. Ibid., 82.

55. Ibid., 108–112.

56. Ibid., 82.

57. Bronn, *Entwickelungs-Gesetze*, 77–78.

58. Junker, "Bronn und *Enstehung der Arten*," 189.

59. Degen, "Vor hundert Jahren"; Gregory, *Scientific Materialism*.

60. Bronn, *Entwickelungs-Gesetze*, 489–490.

61. Ibid., 490.

62. Ibid., 499–500.

3 Darwin's *Origin*

1. Adrian Desmond, "Robert E. Grant: The social predicament of a pre-Darwinian transmutationist," *Journal of the History of Biology* 17 (1984): 189–224; P. Helvig Jespersen, "Charles Darwin and Dr. Grant," *Lychnos* (1948–1949): 159–167; Phillip R. Sloan, "Darwin's invertebrate program, 1826–1836: Preconditions for transformism," in *The Darwinian Heritage*, ed. David Kohn (Princeton: Princeton University Press, 1985), 71–120.

2. Charles Darwin, *The Autobiography of Charles Darwin, 1809–188: With original omissions restored*, ed. Nora Barlow (New York and London: W. W. Norton, 1958), 46–56; Georges Cuvier, *Essay on the Theory of the Earth*, trans. Robert Kerr, with mineralogical notes and an account of Cuvier's geological discoveries by Robert Jameson (Edinburgh: William Blackwood, 1813).

3. Dov Ospovat, *The Development of Darwin's Theory: Natural History, Natural Theology, and Natural Selection, 1838–1859* (Cambridge: Cambridge University Press, 1981; reprint, 1995); Robert M. Young, "Darwin's metaphor: Does nature select?" in *Darwin's Metaphor: Nature's Place in Victorian Culture* (Cambridge: Cambridge University Press, 1985), 79–125; Robert J. Richards, "The theological foundations of Darwin's theory of evolution," in *Experiencing Nature: Proceedings of a Conference in Honor of Allen G. Debus*, ed. Paul Theerman and Karen Parshall (Dordrecht and Boston: Kluwer, 1997), 61–79; Walter F. Cannon, "The bases of Darwin's achievement."

4. Cannon, "Humboldtian science"; Sandra Herbert, *Charles Darwin, Geologist* (Ithaca: Cornell University Press, 2005).

5. Adrian Desmond, *Archetypes and Ancestors: Palaeontology in Victorian London, 1850–1875* (Chicago: University of Chicago Press, 1984); Nicolaas A. Rupke, *Richard Owen: Victorian Naturalist* (New Haven: Yale University Press, 1994).

6. Ospovat, *Development of Darwin's Theory*; Sloan, "Darwin, vital matter, and the transformation of species."

7. Secord, "Nature's fancy"; Secord, "Darwin and the breeders"; Sylvan S. Schweber, "The wider context in Darwin's theorizing," in *The Darwinian Heritage*, ed. David Kohn (Princeton: Princeton University Press, 1985), 35–69; Rebecca Stott, *Darwin and the Barnacle: The Story of One Tiny Creature and History's Most Spectacular Scientific Breakthrough* (London: Faber and Faber, 2003).

8. Darwin, *Autobiography*, 59.

9. William Paley, *Natural Theology; or, Evidences of the Existence and Attributes of the Deity*, 12th ed. (London: J. Faulder, 1809; online version, University of Michigan Humanities Text Initiative, 1998), 1–3, quote on 1. Available from http://quod.lib.umich.edu:80/g/genpub.

10. Ibid., 17–18.

11. Cannon, "Bases of Darwin's achievement"; Ospovat, *Development of Darwin's Theory*; Robert M. Young, "Malthus and the evolutionists: The common context of biological and social theory," in *Darwin's Metaphor: Nature's Place in Victorian Culture* (Cambridge: Cambridge University Press, 1985), 23–55; Young, "Darwin's metaphor"; Elliott Sober, *The Nature of Selection* (Cambridge, Mass.: MIT Press, 1982), ch. 2; Richards, "Theological foundations of Darwin's theory"; Gliboff, "Paley's design argument."

12. Paley, "Natural Theology," 6.

13. Ibid., 60–62.

14. Ibid., 63; on Buffon, 427–430.

15. Ibid., 64–65.

16. Ibid., 64–66

17. Ibid., 72.

18. Ibid., 63.

19. Ibid., 516–522, 531–532.

20. Ibid., 67–68.

21. Ibid., 6.

22. Ibid.

23. Ibid., 7.

24. Ibid., 431–432.

25. Ibid., 476.

26. Ibid., 479–480.

27. Darwin, *Origin of Species*, 1.

28. Ibid., 3.

29. Ibid.

30. Chambers, *Vestiges*, 203–211.

31. Darwin, *Origin of Species*, 3–4.

32 Ibid., 112.

33. *Concordance*, s.v. "improve," "improved" (both counted together), and "perfected." "Perfect" is not used as a verb, in parallel to "improve," and so was not counted.

34. M. J. S. Hodge, "Darwin's argument in the *Origin*," *Philosophy of Science* 59 (1992): 461–464; Ruse, *Darwinian Revolution*; Doren A. Recker, "Causal efficacy: The structure of Darwin's argument strategy in the *Origin of Species*," *Philosophy of Science* 54 (1987): 147–175.

35. Isaac Newton, *Newton's* Principia: *The Mathematical Principles of Natural Philosophy*, trans. Andrew Motte (New York: Daniel Adee, 1846; online version), 384, available from http://www.openlibrary.org/details/100878576.

36. Darwin, *Origin of Species*, 30–31.

37. Letter of 17 September [1861] to Asa Gray, *Darwin Correspondence*, 9: 267.

38. Charles Darwin, *The Variation of Animals and Plants Under Domestication*, 2nd ed., 2 vols. (London: John Murray, 1875; online version), 2: 236, available from http://darwin-online.org.uk/.

39. Ibid.

40. Darwin, *Origin of Species*, 30.

41. Ibid., 55.

42. Ibid.

43. Ibid., 56.

44. Ibid., 62.

45. Ibid., 66–67. In the second edition, Darwin deleted the sentence referring to the wedges in the face of nature, so it did not appear in Bronn's translation. Bronn did, however, read it for his book review, where he took Darwin's metaphors of competition for space a bit too literally.

46. Ibid., 73.

47. Ibid., 79.

48. For further discussion of these points, see Young, "Darwin's metaphor."

49. Darwin, *Origin of Species*, 83.

50. Ibid.

51. Ibid.

52. Ibid., 84

53. Ibid.

54. Ibid., 350.

55. Ibid., 350–351.

56. Ibid., 398–399.

57. Ospovat, *Development of Darwin's Theory*, 210–223.

58. Francis Darwin, ed., *The Foundations of the Origin of Species. Two Essays written in 1842 and 1844 by Charles Darwin* (Cambridge: Cambridge University Press, 1909), 219.

59. Ibid.

60. Ibid.

61. Ospovat, *Development of Darwin's Theory*, 90–101.

62. Darwin, *Origin of Species*, 13–14. This is reiterated in the chapter on morphology, on 444.

63. Ospovat, *Development of Darwin's Theory*, 220–221; see also Darwin, *Origin of Species*, 439–450.

64. F. Darwin, ed., *Foundations of the Origin*, 228–229; Darwin, *Origin of Species*, 440.

65. Darwin, *Origin of Species*, 444.

66. F. Darwin, ed., *Foundations of the Origin*, 230.

67. Richards, *Meaning of Evolution*, 126.

68. F. Darwin, ed., *Foundations of the Origin*, 230.

69. Darwin, *Origin of Species*, 450.

70. Letter of 10 December 1859 to Charles Lyell, *Darwin Correspondence*, 7: 421–424, on 423.

71. Unger, *Botanische Briefe*, 4; Franz Unger, "Steiermark zur Zeit der Braunkohlenbildung," in *Das Alter der Menschheit und das Paradies. Zwei Vorträge,* ed. Franz Unger and [Eduard] Oscar Schmidt (Vienna: Wilhelm Braumüller, 1866).

72. Gliboff, "Evolution, revolution." Unger's opponent, however, did not like natural selection any better than developmental law: Sebastian Brunner, *Das Buch der Natur, mit oder ohne Verfasser? Aphorismen zur Beleuchtung der Darwinslehre* (Vienna: F. Eipeldauer, 1879).

73. Albert Kölliker, "Ueber die Darwin'sche Schöpfungstheorie," *Zeitschrift für wissenschaftliche Zoologie* 14 (1864): 174–186; Carl Nägeli, *Mechanisch-physiologische Theorie der Abstammungslehre* (Munich and Leipzig: R. Oldenbourg, 1884); Julius Wiesner, *Elemente der wissenschaftlichen Botanik*, 2nd ed., vol. 3. "Biologie der Pflanzen" (Vienna: Alfred Hölder, 1902), 277, 287; on Wigand: Montgomery, "Evolution and Darwinism," 97ff.

74. Letter of 18 [and 19] February 1860 to Charles Lyell, *Darwin Correspondence* 8: 92–95.

75. Heinrich G. Bronn, "Review of Ch. Darwin: *On the Origin of Species by means of Natural Selection, or the preservation of favoured races in the struggle for life* (502 pp. 8°, London, 1859)," *Neues Jahrbuch für Mineralogie, Geognosie, Geologie und Petrefaktenkunde* (1860): 112–116, on 112.

76. Ibid.

77. Ibid.

78. Heinrich G. Bronn [?], Review of H. G. Bronn, *Der Stufengang des organischen Lebens von den Insel-Felsen des Ozeans bis auf die Festländer* (31 SS. 8°, Stuttgart 1859), *Neues Jahrbuch für Mineralogie, Geognosie, Geologie und Petrefaktenkunde* (1860): 112.

79. Heinrich G. Bronn, Review of *Origin of Species*, 113.

80. Ibid.

81. Ibid., 115.

82. Ibid., 114.

83. On the experiments to which Bronn was referring, see James Strick, *Sparks of Life. Darwinism and the Victorian Debates Over Spontaneous Generation* (Cambridge, Mass.: Harvard University Press, 2000).

84. Bronn, "Review of *Origin of Species*," 115–116.

85. Ibid., 116.

86. Ibid., 112–113.

87. Letter of 14 February 1860 to H. G. Bronn, *Darwin Correspondence* 8: 82.

88. Ibid.

89. *Variorum*, 68, lines 60*.1–4.

90. Letter of 11 March 1862 to H. G. Bronn, *Darwin Correspondence* 10: 111.

91. Letter of 14 February 1860 to H. G. Bronn, 82–83.

4   Bronn's *Origin*

1. Charles Darwin, *Über die Entstehung der Arten im Thier- und Pflanzen-Reich durch natürliche Züchtung, oder Erhaltung der vervollkommneten Rassen*

im Kampfe um's Daseyn, trans. Heinrich G. Bronn, 1st German ed. (Stuttgart: E. Schweizerbart, 1860), 441n.

2. *Variorum*, 757, lines 255–256.

3. Darwin, *Entstehung der Arten* (1st ed.), 493.

4. F. Darwin, ed., *Life and Letters of Charles Darwin*, 2: 70–71.

5. *Variorum*, 757, line 257.

6. Heinrich G. Bronn, "Schlusswort des Übersetzers zur ersten Deutschen Auflage," in Charles Darwin, *Über die Entstehung der Arten im Thier- und Pflanzen-Reich durch natürliche Züchtung, oder Erhaltung der vervollkommneten Rassen im Kampfe um's Daseyn*, 2nd ed., trans. Heinrich G. Bronn (Stuttgart: E. Schweizerbart, 1863), 525–551, on 538.

7. *Variorum*, 71, line 4c. The reference to the "latter chapters" is new in Darwin's third edition and Bronn's second.

8. Charles Darwin, *Über die Entstehung der Arten im Thier- und Pflanzen-Reich durch natürliche Züchtung*, 2nd ed., 11.

9. Letter of 10 November 1866 to J. V. Carus, *Darwin Correspondence*, 14: 382–384. Charles Darwin, *Über die Entstehung der Arten durch natürliche Zuchtwahl, oder die Erhaltung der begünstigten Rassen im Kampfe um's Dasein*, updated and corrected by J. Victor Carus, trans. Heinrich G. Bronn, 3rd ed. (Stuttgart: E. Schweizerbart, 1867).

10. Bronn, "Schlusswort des Übersetzers," 525.

11. Ibid., 527.

12. Ibid., 526.

13. Ibid.

14. Ibid., 528.

15. Ibid., 529.

16. Ibid., 529 (page misnumbered in the volume consulted—should be 530).

17. Ibid., 529, 540, 543–545.

18. Ibid., 533–539; quote on 533.

19. Ibid., 534.

20. Ibid., 545–548, quote on 547.

21. Ibid., 549.

22. Ibid.

23. Ibid., 551.

24. Letter of 14 July 1860 to H. G. Bronn, *Darwin Correspondence* 8: 287–288.

25. Letter of 5 October 1860 to H. G. Bronn, *Darwin Correspondence* 8: 407–408.

26. *Variorum*, 96, line 128.

27. Darwin, *Variation*, 1: 162–163.

28. Darwin, *Entstehung der Arten* (1st ed.), 41n.

29. Ibid., 41.

30. Darwin, *Origin of Species*, 21. (Unchanged in later editions: *Variorum*, 95, line 120.)

31. Letter of 14 February 1860 to H. G. Bronn, *Darwin Correspondence* 8: 82–83.

32. Historical dictionaries have grafting (*Okulieren*) as the first definition of *Veredlung*, and do not have any agricultural usages for *Adelung*: *Grimms Wörterbuch* and *Adelungs Wörterbuch*, s.v. *Adelung, Veredlung*.

33. Darwin, *Entstehung der Arten* (1st ed.), 10n.

34. Ibid., 344, 344n.

35. Bronn, "Schlusswort des Übersetzers," 540.

36. *Variorum*, 758–759, line 268.

37. Junker, "Bronn und *Enstehung der Arten*," 201.

38. Nyhart, *Biology Takes Form*, 111–112.

39. *Variorum*, 200–201, lines 216–217; Darwin, *Entstehung der Arten* (1st ed.), 113.

40. *Variorum*, 182, line 113.

41. Ibid., 547, line 193.1:b.

42. Ibid., 547, line 193.1:c.

43. For an illuminating discussion of Darwin's views on competitive and morphological "highness," see Timothy Shanahan, *The Evolution of Darwinism: Selection, Adaptation, and Progress in Evolutionary Biology* (Cambridge and New York: Cambridge University Press, 2004), 180–185; also Ospovat, *Development of Darwin's Theory*, ch. 9.

44. Letters of 3 June 1869 from Carus to Darwin and 21 June 1869 from Darwin to Carus, as quoted in Junker, "Bronn und *Enstehung der Arten*," 200–201n.

45. *Grimms Wörterbuch* and *Adelungs Wörterbuch*, s.v. *Ursprung, Entstehung*.

46. Junker, "Bronn und *Enstehung der Arten*," 200.

47. Letter of 15 November 1866 from J. V. Carus, *Darwin Correspondence* 14: 388–390.

48. *Variorum*, 750, lines 196, 201–206.

49. Ibid., 753, line 220:b–220:c. (In contrast, the concluding paragraph of the book adds, in the second edition, that life was breathed "by the Creator," and this phrase is not deleted in the third or later, so Bronn still had some grounds for ascribing theism to Darwin: *Variorum*, 759, line 270:b.)

50. Darwin, *Entstehung der Arten* (2nd ed.), 519n., and Bronn, "Schlusswort des Übersetzers," 529n.

51. Bronn, "Schlusswort des Übersetzers," 546.

52. Ibid., 547.

53. Letter of 5 October 1860 to H. G. Bronn, *Darwin Correspondence* 8: 407–408.

54. *Variorum*, 751, line 206:c.

55. Letter of 5 October 1860 to H. G. Bronn, *Darwin Correspondence* 8: 407–408.

56. *Variorum*, 751, line 206.1:c.

57. Letter of 23 February 1860 to Charles Lyell, *Darwin Correspondence* 8: 102–104.

58. Letter of 5 October 1860 to H. G. Bronn, *Darwin Correspondence* 8: 407–408.

59. Letter of 13 [or 15] October 1860 from H. G. Bronn, *Darwin Correspondence* 8: 425–428.

60. *Variorum*, 495, line 124. In a footnote to Darwin's main text, Bronn gives more precise numbers but does not otherwise contest what Darwin attributes to him: Darwin, *Entstehung der Arten* (2nd ed.), 300n.

61. Bronn, "Schlusswort des Übersetzers," 537.

62. Ibid., 534–536.

63. Ibid., 536.

64. Ibid., 538.

65. Letter of 5 October 1860 to H. G. Bronn, *Darwin Correspondence* 8: 407–408.

66. Letter of 22 May 1863 to George Bentham, *Darwin Correspondence* 11: 432–433.

67. Letter of 11 June 1875 to Lawson Tait, in Francis Darwin and A. C. Seward, eds., *More Letters of Charles Darwin: A Record of His Work in a Series of Hitherto Unpublished Letters*, 2 vols. (New York: Appleton, 1903), 1: 358–359.

68. Letter of 11 May 1880 to T. H. Huxley, in *More Letters of Charles Darwin*, 1: 386–387.

69. Bronn, "Schlusswort des Übersetzers," 544–545.

70. Ibid., 545.

71. Letter of [before 11] March 1862 from H. G. Bronn, *Darwin Correspondence* 10: 110.

5 Ernst Haeckel as a Darwinian Reformer

1. Wilhelm Bölsche, *Ernst Haeckel. Ein Lebensbild*, 2nd ed. (Berlin and Leipzig: Hermann Seemann Nachfolger, n.d., ca. 1900), 104.

2. Ibid., 104–105.

3. Otto Volger, "Über die Darwin'sche Hypothese vom erdwissenschaftlichen Standpunkte aus," *Gesellschaft deutscher Naturforscher und Ärzte, amtlicher Bericht* 38 (1863): 59–72.

4. Ernst Haeckel in *Bericht über die Feier des sechzigsten Geburtstages von Ernst Haeckel am 17. Februar 1894 in Jena* (Jena: privately printed, 1894), 15. On the

chambermaid, see Heinrich Schmidt, "Einleitung," in Ernst Haeckel, *Italienfahrt: Briefe an die Braut, 1859/1860* (Leipzig: K. F. Koehler, 1921), v–viii, on vi.

5. Unless otherwise noted, biographic information is from: Georg Uschmann, *Ernst Haeckel, Forscher, Künstler, Mensch: Briefe, ausgewählt und erläutert* (Leipzig, Jena, and Berlin: Urania-Verlag, 1961) and Erika Krauße, *Ernst Haeckel* (Leipzig: BSB B. G. Teubner, 1984). For more on Gegenbaur's recruiting efforts: Di Gregorio, *Here to Eternity* 51–65. See also Georg Uschmann, *Geschichte der Zoologie und der zoologischen Anstalten in Jena 1771–1919* (Jena: VEB Gustav Fischer Verlag, 1959), 34–40.

6. Ernst Haeckel, *Die Radiolarien (Rhizopoda Radiaria): Eine Monographie.* (Berlin: Georg Reimer, 1862), 1: 231–232, n. 1, quote on 232n.

7. Ibid.

8. Ibid., 1: 231–232, main text.

9. Ibid., 1: 232.

10. Ibid.

11. Ibid., 1: 232–233.

12. Uschmann, *Geschichte der Zoologie in Jena*, 41–46.

13. Coleman, "Between type concept and descent theory"; Erika Krauße, "Zum Verhältnis von Carl Gegenbaur (1826–1903) und Ernst Haeckel (1834–1919)," in *Miscellen zur Geschichte der Biologie: Ilse Jahn und Hans Querner zum 70. Geburtstag*, ed. Armin Geus, Wolfgang Gutmann and Michael Weingarten (Frankfurt: Waldemar Kramer, 1994), 83–89. Di Gregorio makes the relationship much less unequal, but still has Gegenbaur as the dominant partner: Di Gregorio, *Here to Eternity*, 19–21, 81, 156–166, 545–546.

14. Carl Gegenbaur, *Grundzüge der vergleichenden Anatomie* (Leipzig: Wilhelm Engelmann, 1859).

15. Coleman, "Between type concept and descent theory," 171–172; Peter J. Bowler, *Life's Splendid Drama: Evolutionary Biology and the Reconstruction of Life's Ancestry, 1860–1940* (Chicago and London: University of Chicago Press, 1996), 3, 15–18; Di Gregorio, *Here to Eternity*, 155.

16. Di Gregorio, *Here to Eternity*, 78.

17. Darwin, *Origin*, 206; Darwin, *Variorum*, 378, line 266.

18. Darwin, *Origin*, 206; Darwin *Variorum*, 379, line 271.

19. Darwin, *Origin*, 180; Darwin, *Variorum*, 329–331.

20. Andreas Daum, *Wissenschaftspopularisierung im 19. Jahrhundert: Bürgerliche Kultur, naturwissenschaftliche Bildung und die deutsche Öffentlichkeit, 1848–1914* (Munich: R. Oldenbourg, 1998), 119–129; R. Hinton Thomas, *Liberalism, Nationalism, and the German Intellectuals (1822–1847): An Analysis of the Academic and Scientific Conferences of the Period* (Cambridge: W. Heffer and Sons, 1951).

21. Ernst Haeckel, "Über die Entwicklungstheorie Darwins," *Gesellschaft deutscher Naturforscher und Ärzte, amtlicher Bericht* 38 (1863): 16–30, on 16.

22. Ibid., 17.

23. Ibid.

24. Ibid., 21.

25. Krauße, *Ernst Haeckel*, 47–48; Dietmar Schilling, "Biographische und problemgeschichtliche Einleitung," in Jean-Baptiste de Lamarck, *Zoologische Philosophie* (Leipzig: Akademische Verlagsgesellschaft Geest und Portig, 1990).

26. Haeckel, "Entwicklungstheorie," 18.

27. Ibid.

28. Ibid., 18.

29. Ibid.

30. Ibid., 20; cf. Ernst Haeckel, "Ueber die Entwickelungstheorie Darwin's," in *Gesammelte populäre Vorträge aus dem Gebiete der Entwickelungslehre* (Bonn: Emil Strauß, 1878), 1: 1–28, on 10.

31. Haeckel, "Entwicklungstheorie," 22–23.

32. Ibid., 27.

33. Ibid., 22–27; quote on 22.

34. Ibid., 28.

35. Ibid.; see also Haeckel's exchange with the geologist Volger at the Stettin conference, included in Volger, "Darwin'sche Hypothese."

36. Letter of 14 July 1860 to Anna Sethe, in Uschmann, *Forscher, Künstler, Mensch*, 64–67.

37. Ibid.

38. Robert J. Richards, "If this be heresy: Haeckel's conversion to Darwinism," in *Darwinian Heresies*, ed. Abigail Lustig, Robert J. Richards, and Michael Ruse (Cambridge and New York: Cambridge University Press, 2004), 101–130; Di Gregorio, *Here to Eternity*, 20.

39. From Haeckel's speech, as reported in *Bericht über die Feier des sechzigsten Geburtstages*, 19.

40. A descriptor that I am borrowing from Gould, who used it for a different version of Darwinism: Stephen J. Gould, "The hardening of the modern synthesis," in *Dimensions of Darwinism*, ed. Marjorie Grene (Cambridge and New York: Cambridge University Press, and Paris: Editions de la Maison des Sciences de l'Homme, 1983), 71–93.

41. Letter of 13 February 1865 to his parents, in Uschmann, *Forscher, Künstler, Mensch*, 76.

42. Ernst Haeckel, "Fünfzig Jahre Stammesgeschichte: Historisch-kritische Studien über die Resultate der Phylogenie," *Jenaische Zeitschrift für Naturwissenschaft* 54 (= NF 47) (1917): 134–202, on 135.

43. E.g., Ernst Haeckel, *Natürliche Schöpfungs-Geschichte: Gemeinverständliche wissenschaftliche Vorträge über die Entwicklungslehre im allgemeinen und*

diejenige von Darwin, Goethe und Lamarck im besonderen, 11th ed. (Berlin: Georg Reimer, 1911).

44. Ernst Haeckel, *Generelle Morphologie der Organismen: Allgemeine Grundzüge der organischen Formen-Wissenschaft, mechanisch begründet durch die von Charles Darwin reformierte Descendenz-Theorie*, 2 vols. (Berlin: Georg Reimer, 1866), 1: xi–xv.

45. Ibid., 1: 77, 97.

46. Ibid., 1: 100.

47. Ibid., 2: 148–150, on 150.

48. Ibid., 2: 177–180, on 178–179.

49. Ibid.

50. Darwin, *Origin of Species*, e.g., 131–132.

51. Haeckel, *Generelle Morphologie*, 2: 191–193, on 192.

52. Ibid., 2: 180, 186.

53. Ibid., 2: 7–12.

54. Ibid., 2: 190.

55. Darwin, *Origin of Species*, 1314; Darwin, *Entstehung der Arten*, 2nd ed., 24.

56. Haeckel, *Generelle Morphologie*, 2: 184–185, 190.

57. Gould, *Ontogeny and Phylogeny*, 74.

58. Haeckel, *Generelle Morphologie*, 2: 184.

59. Ibid., 2: 185.

60. Ernst Haeckel, *Die Kalkschwämme: Eine Monographie*, 4 vols., vol. 1, Biologie der Kalkschwämme (Calicispongien oder Grantien) (Berlin: Georg Reimer, 1872), 483.

61. A full professorship in zoology had been created for him in 1865, but without the same voting rights in the Philosophical Faculty as enjoyed by the occupants of the original chairs. He accepted the position, but did not give the obligatory inaugural lecture until the restrictions were lifted in 1869: Uschmann, *Geschichte der Zoologie in Jena*, 50–51.

62. Ernst Haeckel, "Ueber Entwicklungsgang und Aufgabe der Zoologie," *Jenaische Zeitschrift für Medicin und Naturwissenschaft* 5 (1870): 353–370, on 370.

63. Ernst Haeckel, "Darwin as an anthropologist," in *Darwin and Modern Science*, ed. A. C. Seward (Cambridge: Cambridge University Press, 1909), 137–151, on 139. See also Haeckel, "Fünfzig Jahre Stammesgeschichte."

64. Haeckel, *Natürliche Schöpfungs-Geschichte*, 191–192, 213.

65. Gavin de Beer, *Embryology and Evolution* (Oxford: Clarendon Press, 1930); Gavin de Beer, *Embryos and Ancestors*, 3rd ed. (Oxford: Clarendon Press, 1958); Gould, *Ontogeny and Phylogeny*, 32–63, 184–186; Bowler, *Non-Darwinian Revolution*, 11, 58, 84.

66. Nicolas Rasmussen also rejects the Baer-Haeckel dichotomy and investigates the motives of de Beer and other twentieth-century anti-Haeckelians, who wanted to reverse the causal arrows and make ontogeny drive phylogeny: Nicolas Rasmussen, "The decline of recapitulationism in early twentieth-century biology: Disciplinary conflict and consensus on the battleground of theory," *Journal of the History of Biology* 24 (1991): 51–89.

67. Letter from Baer to Haeckel, 1876, as quoted in Victor Franz, *Das heutige geschichtliche Bild von Ernst Haeckel* (Jena: Gustav Fischer, 1934), 7.

68. Haeckel, *Generelle Morphologie*, 2: 10–11.

69. Ibid., 2: 259.

70. Ibid., 2: 169.

71. Ibid., 2: 257–258.

72. Ernst Haeckel, "Ueber Arbeitstheilung in Natur- und Menschenleben," in *Gesammelte populäre Vorträge aus dem Gebiete der Entwickelungslehre* (Bonn: Emil Strauß, 1878), 1: 99–141, on 101 (reprint of a lecture from 1868).

73. Haeckel, *Generelle Morphologie*, 2: 261.

74. Ibid.; see also 2: 249–256.

75. Ernst Haeckel, "Die Gastraea-Theorie, die phylogenetische Classification des Thierreichs und die Homologie der Keimblätter," *Jenaische Zeitschrift für Naturwissenschaft* 8 (= NF 1) (1874): 1–55, on 47.

76. Ibid.; Ernst Haeckel, "Die Gastrula und die Eifurchung der Thiere (Fortsetzung der 'Gastraea-Theorie')," *Jenaische Zeitschrift für Naturwissenschaft* 9 (=NF 2) (1875): 402–508 + plates 19–25. For a scientific critique of the Gastraea theory and an account of other attempts to reconstruct the ancestral metazoan, see Karl G. Grell, "Die Gastraea-Theorie," *Medizinhistorisches Journal* 14 (1979): 275–291.

77. Haeckel, *Kalkschwämme*; Ernst Haeckel, "Die Physemarien (Haliphysema und Gastrophysema), Gastraeaden der Gegenwart (Fortsetzung der 'Gastraea-Theorie')," *Jenaische Zeitschrift für Naturwissenschaft* 11 (= NF 4) (1877): 1–54 + plates 1–6; Ernst Haeckel, "Nachträge zur Gastraea-Theorie (Schluss der 'Gastraea-Theorie')," *Jenaische Zeitschrift für Naturwissenschaft* 11 [= NF 4] (1877): 55–98. E. S. Russell pokes fun at Haeckel for the near-Gastraea sponges, which he says had already been recognized and reclassified as foraminifera: Russell, *Form and Function*, 293.

78. Haeckel, "Gastraea-Theorie," 47–48.

Conclusion

1. Gould, *Ontogeny and Phylogeny*, 189–196; Jane Maienschein, "Arguments for experimentation in biology," *PSA—Proceedings of the Biennial Meeting of the Philosophy of Science Association* (1986), 2: 180–195; Jane Maienschein, "Shifting assumptions in American biology: Embryology, 1890–1910," *Journal of the History of Biology* 14 (1981): 89–113; Nick Hopwood, " 'Giving body'

to embryos: Modeling, mechanism, and the microtome in late nineteenth-century anatomy," *Isis* 90 (1999): 462–496.

2. Wilhelm His, "On the principles of animal morphology," *Proceedings of the Royal Society of Edinburgh* 15 (1888): 287–298, on 297; Wilhelm His, *Über die Bedeutung der Entwicklungsgeschichte für die Auffassung der organischen Natur* (Leipzig, 1870), 35, as quoted in Haeckel, "Gastraea-Theorie," 7n.

3. Wilhelm His, *Unsere Körperform und das physiologische Problem ihrer Entstehung* (Leipzig: F. C. W. Vogel, 1874), 159–166. In later works, His was more conciliatory to the evolutionary morphologists and claimed he never denied the importance of heredity and phylogeny. Still, he never changed his definition of heredity and phylogeny. Successive generations of organisms were always periodical consequences of underlying mechanical laws, not of preceding generations: His, "Principles of animal morphology"; Wilhelm His, "Ueber mechanische Grundvorgänge thierischer Formenbildung," *Archiv für Anatomie und Physiologie, anatomische Abteilung* (1894): 1–80.

4. Haeckel, "Gastraea-Theorie," 6.

5. Ibid., 6.

6. Frederick B. Churchill, "Weismann's continuity of the germ-plasm in historical perspective," *Freiburger Universitätsblätter* (1985), no. 87/88: 107–124. See also Frederick B. Churchill, "August Weismann and a break from tradition," *Journal of the History of Biology* 1 (1968): 91–112; Frederick B. Churchill, "Weismann, Hydromedusae, and the biogenetic imperative: A reconsideration," in *A History of Embryology*, ed. T. J. Horder, J. A. Witkowsky and C. C. Wylie (Cambridge: Cambridge University Press, 1986), 7–33.

7. Rasmus G. Winther, "August Weismann on germ-plasm variation," *Journal of the History of Biology* 34 (2001): 517–555.

8. August Weismann, "Zur Frage nach der Vererbung erworbener Eigenschaften," *Biologisches Centralblatt* 6 (1886): 33–48, on 37–38.

9. Ernst Haeckel, "Zur Phylogenie der australischen Fauna: Systematische Einleitung," in *Zoologische Forschungsreisen in Australien und dem malayischen Archipel*, ed. Richard Semon (Jena: Gustav Fischer, 1893), vol. 1, no. 1: i–xxiv, ix–xi.

10. H. Graham Cannon, *Lamarck and Modern Genetics* (Manchester: Manchester University Press, 1959); [Gustav Heinrich] Theodor Eimer, *Die Entstehung der Arten auf Grund von Vererben erworbener Eigenschaften, nach den Gesetzen organischen Wachsens*, 3 vols. (Jena: Gustav Fischer, 1888–1901); Edward D. Cope, *The Primary Factors of Organic Evolution* (Chicago: Open Court, 1896); August Pauly, *Darwinismus und Lamarckismus: Entwurf einer psychophysischen Teleologie* (Munich: Ernst Reinhardt, 1905). On the biometricians and their saltationist rivals, see William B. Provine, *The Origins of Theoretical Population Genetics* (Chicago and London: University of Chicago Press, 1971).

11. Ludwig Plate, *Die Abstammungslehre: Tatsachen, Theorien, Einwände und Folgerungen in kurzer Darstellung*, 2nd ed. (Jena: Gustav Fischer, 1925), 119. Plate counts himself, Roux, Semon, and Haeckel as the principal *Altdarwinisten*.

12. In the British context, George Romanes made a comparable distinction among the overly strict selectionism of Wallace and Weismann, for which he coined the term "neo-Darwinism"; American neo-Lamarckism, which was not selectionistic enough; and his own, more authentic and pluralistic intepretation, for which he had no special name. He did not say where he thought Haeckel fit in this scheme: George Romanes, *Darwin and After Darwin: An Exposition of the Darwinian theory and a discussion of Post-Darwinian Questions*, 2nd ed., 3 vols. (Chicago: Open Court, 1896–1897), 2: 1–20.

13. Haeckel, *Natürliche Schöpfungs-Geschichte*, 205.

14. Haeckel, "Phylogenie der australischen Fauna," vi.

15. Letter of 22 January 1893 to Huxley, in Georg Uschmann and Ilse Jahn, "Der Briefwechsel zwischen Thomas Henry Huxley und Ernst Haeckel: ein Beitrag zum Darwin-Jahr," *Wissenschaftliche Zeitschrift der Friedrich-Schiller-Universität Jena, mathematisch-naturwissenschaftliche Reihe* 9 (1959–60): 7–33, on 27.

16. Richard Semon, *Die Mneme als erhaltendes Prinzip im Wechsel des organischen Geschehens*, 1st and 2nd eds. (Leipzig: Wilhelm Engelmann, 1904, 1908); Richard Semon, "Die somatogene Vererbung im Lichte der Bastard- und Variationsforschung," *Verhandlungen des naturforschenden Vereines in Brünn* 49 (1911): 241–265; Richard Semon, *Das Problem der Vererbung "erworbener Eigenschaften"* (Leipzig: Wilhelm Engelmann, 1912).

17. Ludwig Plate, *Über die Bedeutung des Darwin'schen Selectionsprinzip und Probleme der Artbildung*, 2nd ed. (Leipzig: Wilhelm Engelmann, 1903); Ludwig Plate, *Selektionsprinzip und Probleme der Artbildung: Ein Handbuch des Darwinismus*, 4th ed. (Leipzig and Berlin: Wilhelm Engelmann, 1913); Ludwig Plate, *Vererbungslehre: Mit besonderer Berücksichtigung des Menschen, für Studierende, Ärzte und Züchter*, 1st ed., 2 vols. (Jena: Gustav Fischer, 1913); Georgy S. Levit and Uwe Hoßfeld, "The forgotten 'old-Darwinian' synthesis: The evolutionary theory of Ludwig H. Plate (1862–1937)," *NTM—Internationale Zeitschrift für Geschichte und Ethik der Naturwissenschaften, Technik und Medizin* 14 (2006): 9–25.

18. Paul Kammerer, "Mendelsche Regeln und Vererbung erworbener Eigenschaften," *Verhandlungen des naturforschenden Vereines in Brünn* 49 (1911): 72–110; Paul Kammerer, "Adaptation and inheritance in the light of modern experimental investigation," *Annual Report of the Board of Regents of the Smithsonian Institution* (1912): 421–441; Paul Kammerer, "Haeckel und ich; Der Planet und der Kieselstein," in *Was wir Ernst Haeckel verdanken*, ed. Heinrich Schmidt (Leipzig: Unesma, 1914), 2: 6–14; Paul Kammerer, *The Inheritance of Acquired Characteristics*, trans. A. Paul Maerker-Branden (New York: Boni and Liveright, 1924); Sander Gliboff, "'Protoplasm . . . is soft wax in our hands': Paul Kammerer and the art of animal transformation," *Endeavour* 29 (2005): 162–165; Sander Gliboff, "The case of Paul Kammerer: Evolution and experimentation in the early twentieth century," *Journal of the History of Biology* 39 (2006): 525–563.

19. Peter J. Bowler, *The Eclipse of Darwinism: Anti-Darwinian Theories in the Decades around 1900* (Baltimore and London: Johns Hopkins University Press, 1983); Julian S. Huxley, *Evolution: The Modern Synthesis* (New York and London: Harper, 1942), 22–28.

20. William B. Provine, "Progress in evolution and meaning in life," in *Julian Huxley: Biologist and Statesman of Science*, ed. C. Kenneth Waters and Albert van Helden (Houston: Rice University Press, 1992), 165–180.

21. Hull, "Darwinism as a historical entity."

22. Giuliano Pancaldi, *Darwin in Italy: Science Across Cultural Frontiers*, trans. Ruey Morelli, 2nd ed. (Bloomington and Indianapolis: Indiana University Press, 1991); James Pusey, *China and Charles Darwin* (Cambridge, Mass.: Harvard University Press, 1983).

23. Daniel P. Todes, *Darwin without Malthus: The Struggle for Existence in Russian Evolutionary Thought* (New York and Oxford: Oxford University Press, 1989).

24. Darwin, *Autobiography*, 120.

Bibliography

Reference Works

ADB = *Allgemeine deutsche Biographie*. 56 vols. Leipzig: Duncker and Humblot, 1875–1912.

Adelungs Wörterbuch = *Grammatisch-kritisches Wörterbuch*. 2nd ed. 4 vols. Edited by Johann C. Adelung. Leipzig: J. G. I. Breitkopf, 1793–1801.

Concordance = *A Concordance to Darwin's Origin of Species*. Edited by Paul H. Barrett, Donald J. Weinshank, and Timothy T. Gottleber. Ithaca, N.Y.: Cornell University Press, 1981.

Darwin Correspondence = *The Correspondence of Charles Darwin*. Edited by Frederick Burkhardt, Sydney Smith, et al. Cambridge and New York: Cambridge University Press, 1985–.

DSB = *Dictionary of Scientific Biography*. Edited by Charles C. Gillispie, et al. New York: Charles Scribner's Sons, 1970–1981.

Marginalia = *Charles Darwin's Marginalia*. Vol. 1. Edited by Mario Di Gregorio. New York and London: Garland, 1990.

NDB = *Neue deutsche Biographie*. Berlin: Duncker and Humblot, 1953–.

Grimms Wörterbuch = *Deutsches Wörterbuch*. Edited by Jacob Grimm, et al. Leipzig: S. Hirzel, 1854–1971.

Variorum = *The Origin of Species by Charles Darwin: A Variorum Text*. Edited by Morse Peckham. Philadelphia: University of Pennsylvania Press, 1959.

Primary Sources

Baer, Karl Ernst von. *Über Entwickelungsgeschichte der Thiere: Beobachtung und Reflexion*. 2 vols. Königsberg: Bornträger, 1828–1837.

Bericht über die Feier des sechzigsten Geburtstages von Ernst Haeckel am 17. Februar 1894 in Jena. Jena: privately printed, 1894.

Blumenbach, Johann F. *Über den Bildungstrieb und das Zeugungsgeschäfte*. 1st ed. Göttingen: Johann Christian Dieterich, 1781; facsimile, Stuttgart: Gustav Fischer, 1971.

Blumenbach, Johann F. *Über den Bildungstrieb und das Zeugungsgeschäfte.* 2nd ed. Göttingen: Johann Christian Dieterich, 1791.

Blumenbach, Johann F. *Handbuch der Naturgeschichte.* 11th ed. Göttingen: Dieterich, 1825.

Bronn, Heinrich G. *Handbuch einer Geschichte der Natur.* 3 vols. with atlas. Stuttgart: E. Schweizerbart, 1841–1849.

Bronn, Heinrich G. *Untersuchungen über die Entwickelungs-Gesetze der organischen Welt während der Bildungs-Zeit unserer Erd-Oberfläche: Eine von der Französischen Akademie im Jahre 1857 gekrönte Preisschrift.* Stuttgart: E. Schweizerbart, 1858.

Bronn, Heinrich G. *Morphologische Studien über die Gestaltungs-Gesetze der Naturkörper überhaupt und der organischen insbesondere.* Leipzig and Heidelberg: C. F. Winter, 1858.

Bronn, Heinrich G. "Review of Ch. Darwin, *On the Origin of Species by means of Natural Selection, or the preservation of favoured races in the struggle for life* (502 pp. 8°, London, 1859)." *Neues Jahrbuch für Mineralogie, Geognosie, Geologie und Petrefaktenkunde* (1860): 112–116.

Bronn, Heinrich G. [?]. "Review of H. G. Bronn, *Der Stufengang des organischen Lebens von den Insel-Felsen des Ozeans bis auf die Festländer* (31 SS. 8°, Stuttgart 1859)." *Neues Jahrbuch für Mineralogie, Geognosie, Geologie und Petrefaktenkunde* (1860): 112.

Bronn, Heinrich G. "Schlusswort des Übersetzers zur ersten Deutschen Auflage." *Über die Entstehung der Arten im Thier- und Pflanzen-Reich durch natürliche Züchtung, oder Erhaltung der vervollkommneten Rassen im Kampfe um's Daseyn,* 525–551. Stuttgart: E. Schweizerbart, 1863.

Brunner, Sebastian. *Das Buch der Natur, mit oder ohne Verfasser? Aphorismen zur Beleuchtung der Darwinslehre.* Vienna: F. Eipeldauer, 1879.

Burdach, Karl F. *Propädeutik zum Studium der gesammten Heilkunst: Ein Leitfaden akademischer Vorlesungen.* Leipzig: Breitkopf und Härtel, 1800.

Chambers, Robert. *Natürliche Geschichte der Schöpfung des Weltalls, der Erde und der auf ihr befindlichen Organismen, begründet auf die durch die Wissenschaft errungenen Thatsachen.* 2nd ed. Trans. Carl Vogt. Braunschweig: Friedrich Vieweg, 1858.

Chambers, Robert. *Vestiges of the Natural History of Creation.* 1st ed., 1844. Fascimile in James A. Secord, ed., *Vestiges of the Natural History of Creation and Other Evolutionary Writings.* Chicago: University of Chicago Press, 1994.

Cope, Edward D. *The Primary Factors of Organic Evolution.* Chicago: Open Court, 1896.

Cuvier, Georges. *Essay on the Theory of the Earth.* With mineralogical notes and an account of Cuvier's geological discoveries by Robert Jameson. Trans. Robert Kerr. Edinburgh: William Blackwood, 1813.

Cuvier, Georges. *The Animal Kingdom, Arranged After Its Organization, Forming a Natural History of Animals and an Introduction to Comparative Anatomy.* Trans. Edward Blyth et al. London: William S. Orr, 1849.

Darwin, Charles. *On the Origin of Species by Means of Natural Selection, or the Preservation of Favoured Races in the Struggle for Life*. 1st ed. London: John Murray, 1859. Facsimile, Cambridge, Mass.: Harvard University Press, 1964.

Darwin, Charles. *Über die Entstehung der Arten im Thier- und Pflanzen-Reich durch natürliche Züchtung, oder Erhaltung der vervollkommneten Rassen im Kampfe um's Daseyn*. 1st ed. Trans. Heinrich G. Bronn. Stuttgart: E. Schweizerbart, 1860.

Darwin, Charles. *Über die Entstehung der Arten im Thier- und Pflanzen-Reich durch natürliche Züchtung, oder Erhaltung der vervollkommneten Rassen im Kampfe um's Daseyn*. 2nd. ed. Trans. Heinrich G. Bronn. Stuttgart: E. Schweizerbart, 1863.

Darwin, Charles. *Über die Entstehung der Arten durch natürliche Zuchtwahl, oder die Erhaltung der begünstigten Rassen im Kampfe um's Dasein*. 3rd ed. Updated and corrected by J. Victor Carus. Trans. Heinrich G. Bronn. Stuttgart: E. Schweizerbart, 1867.

Darwin, Charles. *The Variation of Animals and Plants Under Domestication*. 2nd ed. 2 vols. London: John Murray, 1875. Online version from http://darwin-online.org.uk.

Darwin, Charles. *The Autobiography of Charles Darwin, 1809–1882: With original omissions restored*. Edited by Nora Barlow. New York and London: W. W. Norton, 1958.

Darwin, Charles, and Alfred R. Wallace. "On the tendency of species to form varieties; and on the perpetuation of varieties and species by natural means of selection." *Journal of the Linnean Society of London* 3 (1858): 45–62.

Darwin, Francis, ed. *The Life and Letters of Charles Darwin: Including an Autobiographical Chapter*. 2 vols. New York: D. Appleton, 1898.

Darwin, Francis, ed. *The Foundations of the Origin of Species: Two Essays written in 1842 and 1844 by Charles Darwin*. Cambridge: Cambridge University Press, 1909.

Darwin, Francis, and A. C. Seward, eds. *More Letters of Charles Darwin: A Record of His Work in a Series of Hitherto Unpublished Letters*. 2 vols. New York: Appleton, 1903.

Eimer, [Gustav Heinrich] Theodor. *Die Entstehung der Arten auf Grund von Vererben erworbener Eigenschaften, nach den Gesetzen organischen Wachsens*. 3 vols. Jena: Gustav Fischer, 1888–1901.

Gegenbaur, Carl. *Grundzüge der vergleichenden Anatomie*. Leipzig: Wilhelm Engelmann, 1859.

Goethe, Johann W. von. *Die Metamorphose der Pflanzen*. Gotha: Carl Wilhelm Ettinger, 1790. Facsimile, Weinheim: Acta Humaniora, 1984.

Haeckel, Ernst. *Die Radiolarien (Rhizopoda Radiaria): Eine Monographie*. With atlas. Berlin: Georg Reimer, 1862.

Haeckel, Ernst. "Über die Entwicklungstheorie Darwins." *Gesellschaft deutscher Naturforscher und Ärzte, amtlicher Bericht* 38 (1863): 17–30.

Haeckel, Ernst. *Generelle Morphologie der Organismen: Allgemeine Grundzüge der organischen Formen-Wissenschaft, mechanisch begründet durch die von Charles Darwin reformierte Descendenz-Theorie.* 2 vols. Berlin: Georg Reimer, 1866.

Haeckel, Ernst. "Ueber Entwicklungsgang und Aufgabe der Zoologie." *Jenaische Zeitschrift für Medicin und Naturwissenschaft* 5 (1870): 353–370.

Haeckel, Ernst. *Die Kalkschwämme: Eine Monographie.* 4 vols. Vol. 1, *Biologie der Kalkschwämme (Calicispongien oder Grantien).* Berlin: Georg Reimer, 1872.

Haeckel, Ernst. "Die Gastraea-Theorie, die phylogenetische Classification des Thierreichs und die Homologie der Keimblätter." *Jenaische Zeitschrift für Naturwissenschaft* 8 (= NF 1) (1874): 1–55.

Haeckel, Ernst. "Die Gastrula und die Eifurchung der Thiere (Fortsetzung der 'Gastraea-Theorie')." *Jenaische Zeitschrift für Naturwissenschaft* 9 (= NF 2) (1875): 402–508 + plates 19–25.

Haeckel, Ernst. "Die Physemarien (Haliphysema und Gastrophysema), Gastraeaden der Gegenwart (Fortsetzung der 'Gastraea-Theorie')." *Jenaische Zeitschrift für Naturwissenschaft* 11 (= NF 4) (1877): 1–54 + plates 1–6.

Haeckel, Ernst. "Nachträge zur Gastraea-Theorie (Schluss der 'Gastraea-Theorie')." *Jenaische Zeitschrift für Naturwissenschaft* 11 (= NF 4) (1877): 55–98.

Haeckel, Ernst. "Ueber die Entwickelungstheorie Darwin's." *Gesammelte populäre Vorträge aus dem Gebiete der Entwickelungslehre* 1: 1–28. Bonn: Emil Strauß, 1878.

Haeckel, Ernst. "Ueber Arbeitstheilung in Natur- und Menschenleben." *Gesammelte populäre Vorträge aus dem Gebiete der Entwickelungslehre* 1: 99–141. Bonn: Emil Strauß, 1878.

Haeckel, Ernst. "Zur Phylogenie der australischen Fauna: Systematische Einleitung." In Richard Semon, ed., *Zoologische Forschungsreisen in Australien und dem malayischen Archipel*, vol. 1, *Ceratodus*, no. 1: i–xxiv. Jena: Gustav Fischer, 1893.

Haeckel, Ernst. *Die Welträthsel: Gemeinverständliche Studien über Monistische Philosophie.* Bonn: Emil Strauß, 1899.

Haeckel, Ernst. *Anthropogenie oder Entwickelungsgeschichte der Menschen.* 5th ed. 2 vols. Leipzig: Wilhelm Engelmann, 1903.

Haeckel, Ernst. "Darwin as an anthropologist." In A. C. Seward, ed., *Darwin and Modern Science*, 137–151. Cambridge: Cambridge University Press, 1909.

Haeckel, Ernst. *Der Monismus als Band zwischen Religion und Wissenschaft: Glaubensbekenntnis eines Naturforschers.* 15th ed. Leipzig: Alfred Kröner Verlag, 1911.

Haeckel, Ernst. *Natürliche Schöpfungs-Geschichte: Gemeinverständliche wissenschaftliche Vorträge über die Entwicklungslehre im allgemeinen und diejenige von Darwin, Goethe und Lamarck im besonderen.* 11th ed. Berlin: Georg Reimer, 1911.

Haeckel, Ernst. "Fünfzig Jahre Stammesgeschichte: Historisch-kritische Studien über die Resultate der Phylogenie." *Jenaische Zeitschrift für Naturwissenschaft* 54 (= NF 47) (1917): 134–202.

Hirschfeld, Magnus. "Ernst Haeckel und die Sexualwissenschaft." In Heinrich Schmidt, ed., *Was wir Ernst Haeckel verdanken: Ein Buch der Verehrung und Dankbarkeit*, 2: 282–284. Leipzig: Unesma, 1914.

His, Wilhelm. *Unsere Körperform und das physiologische Problem ihrer Entstehung*. Leipzig: F. C. W. Vogel, 1874.

His, Wilhelm. "On the principles of animal morphology." *Proceedings of the Royal Society of Edinburgh* 15 (1888): 287–298.

His, Wilhelm. "Ueber mechanische Grundvorgänge thierischer Formenbildung." *Archiv für Anatomie und Physiologie, anatomische Abteilung* (1894): 1–80.

Kammerer, Paul. "Mendelsche Regeln und Vererbung erworbener Eigenschaften." *Verhandlungen des naturforschenden Vereines in Brünn* 49 (1911): 72–110.

Kammerer, Paul. "Adaptation and inheritance in the light of modern experimental investigation." *Annual Report of the Board of Regents of the Smithsonian Institution* (1912): 421–441.

Kammerer, Paul. "Haeckel und ich: Der Planet und der Kieselstein." In Heinrich Schmidt, ed., *Was wir Ernst Haeckel verdanken: Ein Buch der Verehrung und Dankbarkeit*, 2: 6–14. Leipzig: Unesma, 1914.

Kammerer, Paul. *The Inheritance of Acquired Characteristics*. Trans. A. Paul Maerker-Branden. New York: Boni and Liveright, 1924.

Kant, Immanuel. *Kants gesammelte Schriften*. Akademie edition. Berlin: Georg Reimer, 1902–1923. Online version from http://www.ikp.uni-bonn.de/kant/.

Kielmeyer, Carl F. *Ueber die Verhältniße der organischen Kräfte unter einander in der Reihe der verschiedenen Organisationen, die Gesetze und Folgen dieser Verhältniße*. Stuttgart, 1793. Facsimile, Marburg a. d. Lahn: Basilisken-Presse, 1993.

Kölliker, Albert. "Ueber die Darwin'sche Schöpfungstheorie." *Zeitschrift für wissenschaftliche Zoologie* 14 (1864): 174–186.

Lamarck, Jean-Baptiste de. *Hydrogéologie, ou, Recherches sur l'influence qu'ont les eaux sur la surface du globe terrestre*. Paris: Chez l'auteur, 1802.

Meckel, Johann F. "Beyträge zur Geschichte des menschlichen Fötus." *Beyträge zur vergleichenden Anatomie*. Vol. 1, Book 1: 57–124. Leipzig: Carl Heinrich Reclam, 1808.

Meckel, Johann F. "Besprechung dreyer kopfloser Missgeburten, nebst einigen allgemeinen Bemerkungen über diese Art von Missbildung." *Beyträge zur vergleichenden Anatomie*. Vol. 1, Book 2: 136–162. Leipzig: Carl Heinrich Reclam, 1809.

Meckel, Johann F. "Entwurf einer Darstellung der zwischen dem Embryonalzustande der höhern Thiere und dem permanenten der niedern Statt findenden Parallele." *Beyträge zur vergleichenden Anatomie*. Vol. 2, Book 1: 1–60. Leipzig: Carl Heinrich Reclam, 1811.

Meckel, Johann F. "Ueber den Charakter der allmählichen Vervollkommnung der Organisation, oder den Unterschied zwischen den höhern und niedern Bildungen." *Beyträge zur vergleichenden Anatomie.* Vol. 2, Book 1: 61–123. Leipzig: Carl Heinrich Reclam, 1811.

Meckel, Johann F. *System der vergleichenden Anatomie.* 5 vols. Halle: Renger, 1821.

Nägeli, Carl. *Mechanisch-physiologische Theorie der Abstammungslehre.* Munich and Leipzig: R. Oldenbourg, 1884.

Newton, Isaac. *Newton's Principia: The Mathematical Principles of Natural Philosophy,* 1st American ed. Trans. Andrew Mott. New York: Daniel Adee, 1846. Online version from http://www.openlibrary.org/details/100878576.

Oken, Lorenz. *Lehrbuch der Naturphilosophie.* 2nd ed. Jena: Friedrich Frommann, 1831.

Oken, Lorenz, and Dietrich Georg von Kieser, eds. *Beiträge zur vergleichenden Zoologie, Anatomie und Physiologie.* 2 vols. Bamberg: Joseph Anton Göbhardt, 1806–1807.

Paley, William. *Natural Theology; or, Evidences of the Existence and Attributes of the Deity.* 12th ed. London: J. Faulder, 1809. Online version, University of Michigan Humanities Text Initiative, 1998, from http://quod.lib.umich.edu:80/g/genpub.

Pander, Christian. "Allgemeine Bemerkungen über die äußeren Einflüsse auf die organische Entwicklung der Thiere." In Christian Pander and Eduard d'Alton, Sr., eds., *Die Vergleichende Osteologie.* Bonn: Eduard Weber, 1824.

Pauly, August. *Darwinismus und Lamarckismus: Entwurf einer psychophysischen Teleologie.* Munich: Ernst Reinhardt, 1905.

Plate, Ludwig. *Über die Bedeutung des Darwin'schen Selectionsprinzip und Probleme der Artbildung.* 2nd ed. Leipzig: Wilhelm Engelmann, 1903.

Plate, Ludwig. *Selektionsprinzip und Probleme der Artbildung: Ein Handbuch des Darwinismus.* 4th ed. Leipzig and Berlin: Wilhelm Engelmann, 1913.

Plate, Ludwig. *Vererbungslehre: Mit besonderer Berücksichtigung des Menschen, für Studierende, Ärzte und Züchter.* 1st ed., 2 vols. Jena: Gustav Fischer, 1913.

Plate, Ludwig. *Die Abstammungslehre: Tatsachen, Theorien, Einwände und Folgerungen in kurzer Darstellung.* 2nd ed. Jena: Gustav Fischer, 1925.

Reil, Johann C. "Von der Lebenskraft." *Gesammelte kleine physiologische Schriften,* 1: 3–133. Vienna: Aloys Doll, 1811. (Reprinted from *Archiv für Physiologie* 1 [1796]: 8–162.)

Romanes, George J. *Darwin and After Darwin: An Exposition of the Darwinian Theory and a Discussion of Post-Darwinian Questions.* 2nd ed., 3 vols. Chicago: Open Court, 1896–1897.

Schleiden, Matthias. *Die Pflanze und ihr Leben: Populäre Vorträge.* 1st ed. Leipzig: Wilhelm Engelmann, 1848.

Schmidt, Heinrich, ed. *Was wir Ernst Haeckel verdanken: Ein Buch der Verehrung und Dankbarkeit.* 2 vols. Leipzig: Unesma, 1914.

Schmidt, Heinrich. "Einleitung." In Ernst Haeckel, *Italienfahrt: Briefe an die Braut, 1859/1860*, v–viii. Leipzig: K. F. Koehler, 1921.

Semon, Richard. *Die Mneme als erhaltendes Prinzip im Wechsel des organischen Geschehens.* 1st ed. Leipzig: Wilhelm Engelmann, 1904.

Semon, Richard. *Die Mneme als erhaltendes Prinzip im Wechsel des organischen Geschehens.* 2nd ed. Leipzig: Wilhelm Engelmann, 1908.

Semon, Richard. "Die somatogene Vererbung im Lichte der Bastard- und Variationsforschung." *Verhandlungen des naturforschenden Vereines in Brünn* 49 (1911): 241–265.

Semon, Richard. *Das Problem der Vererbung "erworbener Eigenschaften."* Leipzig: Wilhelm Engelmann, 1912.

Stöcker, Helene. [Untitled]. In Heinrich Schmidt, ed., *Was wir Ernst Haeckel verdanken: Ein Buch der Verehrung und Dankbarkeit*, 2: 324–328. Leipzig: Unesma, 1914.

Treviranus, Gottfried R. *Biologie, oder Philosophie der lebenden Natur für Naturforscher und Aerzte.* 6 vols. Göttingen: Johann Friedrich Röwer, 1802–1822.

Unger, Franz. *Botanische Briefe.* Vienna: Carl Gerold und Sohn, 1852.

Unger, Franz. "Steiermark zur Zeit der Braunkohlenbildung." In Franz Unger and [Eduard] Oscar Schmidt, eds., *Das Alter der Menschheit und das Paradies: Zwei Vorträge.* Vienna: Wilhelm Braumüller, 1866.

Unger, Franz. *Versuch einer Geschichte der Pflanzenwelt.* Vienna: Wilhelm Braumüller, 1852.

Uschmann, Georg, and Ilse Jahn. "Der Briefwechsel zwischen Thomas Henry Huxley und Ernst Haeckel: ein Beitrag zum Darwin-Jahr." *Wissenschaftliche Zeitschrift der Friedrich-Schiller-Universität Jena, mathematisch-naturwissenschaftliche Reihe* 9 (1959–60): 7–33.

Uschmann, Georg. *Ernst Haeckel, Forscher, Künstler, Mensch: Briefe, ausgewählt und erläutert.* Leipzig, Jena, and Berlin: Urania-Verlag, 1961.

Volger, Otto. "Über die Darwin'sche Hypothese vom erdwissenschaftlichen Standpunkte aus." *Gesellschaft deutscher Naturforscher und Ärzte, amtlicher Bericht* 38 (1863): 59–72.

Weismann, August. "Zur Frage nach der Vererbung erworbener Eigenschaften." *Biologisches Centralblatt* 6 (1886): 33–48.

Wiesner, Julius. *Elemente der wissenschaftlichen Botanik.* 2nd ed., vol. 3, *Biologie der Pflanzen.* Vienna: Alfred Hölder, 1902.

Secondary Sources

Allen, Ann Taylor. *Feminism and Motherhood in Germany, 1800–1914.* New Brunswick, N.J.: Rutgers University Press, 1991.

Amundson, Ronald. *The Changing Role of the Embryo in Evolutionary Thought: Roots of Evo-Devo.* Cambridge and New York: Cambridge University Press, 2005.

Bach, Thomas. "'Was ist das Thierreich anders als der anatomirte Mensch . . . ?': Oken in Göttingen (1805–1807)." In Olaf Breidbach, Hans-Joachim Fliedner, and Klaus Ries, eds., *Lorenz Oken (1779–1851): Ein politischer Naturphilosoph*, 73–91. Weimar: Hermann Böhlau, 2001.

Baron, Walter. "Zur Stellung von Heinrich Georg Bronn (1800–1862) in der Geschichte des Evolutionsgedankens." *Sudhoffs Archiv für Geschichte der Medizin und der Naturwissenschaften* 45 (1961): 97–109.

Ben-David, Joseph. *The Scientist's Role in Society: A Comparative Study*. Englewood Cliffs, N.J.: Prentice-Hall, 1971.

Beneke, Rudolf. *Johann Friedrich Meckel der Jüngere*. Halle: Max Niemeyer, 1934.

Benton, Ted. "Social Darwinism and Socialist Darwinism in Germany: 1860–1900." *Rivista di Filosofia* 23 (1982): 79–121.

Blackbourn, David, and Geoff Eley. *The Peculiarities of German History: Bourgeois Society and Politics in Nineteenth-Century Germany*. Oxford: Oxford University Press, 1984.

Bolle, Fritz. "Monistische Maurerei." *Medizinhistorisches Journal* 16 (1981): 280–301.

Bölsche, Wilhelm. *Ernst Haeckel: Ein Lebensbild*. 2nd ed. Berlin and Leipzig: Hermann Seemann, n.d., ca. 1900.

Borscheid, Peter. *Naturwissenschaft, Staat und Industrie in Baden (1848–1914)*. Stuttgart: Ernst Klett, 1976.

Bowler, Peter J. *Fossils and Progress: Paleontology and the Idea of Progressive Evolution in the Nineteenth Century*. New York: Science History Publications, 1976.

Bowler, Peter J. *The Eclipse of Darwinism: Anti-Darwinian Theories in the Decades around 1900*. Baltimore: Johns Hopkins University Press, 1983.

Bowler, Peter J. *The Non-Darwinian Revolution: Reinterpreting a Historical Myth*. Baltimore: Johns Hopkins University Press, 1988.

Bowler, Peter J. *Life's Splendid Drama: Evolutionary Biology and the Reconstruction of Life's Ancestry, 1860–1940*. Chicago: University of Chicago Press, 1996.

Breidbach, Olaf. "Darwinism and comparative anatomy in 19th century Germany." In Alessandro Minelli and Sandra Casellato, eds., *Giovanni Canestrini, Zoologist and Darwinist*, 149–167. Venice: Istituto Veneto di Scienze, Lettere ed Arti, 2001.

Broman, Thomas. "University reform in medical thought at the end of the eighteenth century." *Osiris* 2nd series, 5 (1989): 36–53.

Browne, Janet. *The Secular Ark: Studies in the History of Biogeography*. New Haven: Yale University Press, 1983.

Browne, Janet. *Charles Darwin: Voyaging*. London: Jonathan Cape, 1995.

Browne, Janet. *Charles Darwin: The Power of Place*. New York: Alfred A. Knopf, 2003.

Bruford, William H. *The German Tradition of Self-Cultivation: "Bildung" from Humboldt to Thomas Mann.* Cambridge: Cambridge University Press, 1975.

Caneva, Kenneth L. "Teleology with regrets: Review of Timothy Lenoir, *The Strategy of Life."* *Annals of Science* 47 (1990): 291–300.

Cannon, H. Graham. *Lamarck and Modern Genetics.* Manchester: Manchester University Press, 1959.

Cannon, Susan F. "Humboldtian science." In *Science in Culture: The Early Victorian Period,* 73–110. New York: Science History Publications, 1978.

Cannon, Walter F. "The bases of Darwin's achievement: A reevaluation." *Victorian Studies* 5 (1961): 109–134.

Churchill, Frederick B. "August Weismann and a break from tradition." *Journal of the History of Biology* 1 (1968): 91–112.

Churchill, Frederick B. "Weismann's continuity of the germ-plasm in historical perspective." *Freiburger Universitätsblätter* 87/88 (1985): 107–124.

Churchill, Frederick B. "Weismann, Hydromedusae, and the biogenetic imperative: A reconsideration." In T. J. Horder, J. A. Witkowsky, and C. C. Wylie, eds., *A History of Embryology,* 7–33. Cambridge: Cambridge University Press, 1986.

Coleman, William. *Georges Cuvier, Zoologist: A Study in the History of Evolution Theory.* Cambridge, Mass.: Harvard University Press, 1964.

Coleman, William. *Biology in the Nineteenth Century: Problems of Form, Function, and Transformation.* 1971. Reprint, Cambridge: Cambridge University Press, 1977.

Coleman, William. "Limits of the recapitulation theory: Carl Friedrich Kielmeyer's critique of the presumed parallelism of earth history, ontogeny, and the present order of organisms." *Isis* 64 (1973): 341–350.

Coleman, William. "Morphology between type concept and descent theory." *Journal of the History of Medicine and Allied Sciences* 31 (1976): 149–175.

Daum, Andreas. *Wissenschaftspopularisierung im 19. Jahrhundert: Bürgerliche Kultur, naturwissenschaftliche Bildung und die deutsche Öffentlichkeit, 1848–1914.* Munich: R. Oldenbourg, 1998.

Dawkins, Richard. *The Blind Watchmaker: Why the Evidence of Evolution Reveals a Universe without Design.* 1985. Reprint, New York and London: W. W. Norton, 1987.

de Beer, Gavin R. *Embryology and Evolution.* Oxford: Clarendon, 1930.

de Beer, Gavin R. *Embryos and Ancestors.* 3rd ed. Oxford: Clarendon, 1958.

Degen, Heinz. "Vor hundert Jahren: Die Naturforscherversammlung zu Göttingen und der Materialismusstreit." *Naturwissenschaftliche Rundschau* 7 (1954): 271–277.

Desmond, Adrian. *Archetypes and Ancestors: Palaeontology in Victorian London, 1850–1875.* Chicago: University of Chicago Press, 1984.

Desmond, Adrian. "Robert E. Grant: The social predicament of a pre-Darwinian transmutationist." *Journal of the History of Biology* 17 (1984): 189–224.

Desmond, Adrian, and James R. Moore. *Darwin: The Life of a Tormented Evolutionist.* New York and London: W. W. Norton, 1994.

Di Gregorio, Mario. *From Here to Eternity: Ernst Haeckel and Scientific Faith.* Göttingen: Vandenhoeck und Ruprecht, 2005.

Eley, Geoff, ed. *Society, Culture, and the State in Germany, 1870–1930.* Ann Arbor: University of Michigan Press, 1996.

Farber, Paul L. "The type concept in zoology in the first half of the nineteenth century." *Journal of the History of Biology* 9 (1976): 93–119.

Forman, Paul. "Weimar culture, causality, and quantum theory, 1918–1927: Adaptation by German physicists and mathematicians to a hostile intellectual environment." *Historical Studies in the Physical Sciences* 3 (1971): 1–115.

Franz, Victor. *Das heutige geschichtliche Bild von Ernst Haeckel.* Jena: Gustav Fischer, 1934.

Gasman, Daniel. *The Scientific Origins of National Socialism: Social Darwinism in Ernst Haeckel and the Monist League.* New York: American Elsevier, 1971.

Gasman, Daniel. *Haeckel's Monism and the Birth of Fascist Ideology.* New York: Peter Lang, 1997.

Gliboff, Sander. "Evolution, revolution, and reform in Vienna: Franz Unger's ideas on descent and their post-1848 reception." *Journal of the History of Biology* 31 (1998): 179–209.

Gliboff, Sander. "Gregor Mendel and the laws of evolution." *History of Science* 37 (1999): 217–235.

Gliboff, Sander. "Paley's design argument as an inference to the best explanation, or, Dawkins' dilemma." *Studies in the History and Philosophy of Biological and Biomedical Sciences* 31 (2000): 579–597.

Gliboff, Sander. "Darwin on Trial Again: Review of *From Darwin to Hitler. Evolutionary Ethics, Eugenics, and Racism in Germany*, by Richard Weikart (2004)." Electronic publication: H-German, 2004, from http://www.h-net.org/reviews/.

Gliboff, Sander. "'Protoplasm . . . is soft wax in our hands': Paul Kammerer and the art of animal transformation." *Endeavour* 29 (2005): 162–165.

Gliboff, Sander. "The case of Paul Kammerer: Evolution and experimentation in the early twentieth century." *Journal of the History of Biology* 39 (2006): 525–563.

Goldschmidt, Richard. *Portraits from Memory: Recollections of a Zoologist.* Seattle: University of Washington Press, 1956.

Gould, Stephen J. *Ontogeny and Phylogeny.* Cambridge, Mass.: Harvard University/Belknap Press, 1977.

Gould, Stephen J. "The hardening of the modern synthesis." In Marjorie Grene, ed., *Dimensions of Darwinism,* 71–93. Cambridge and New York: Cambridge University Press, and Paris: Editions de la Maison des Sciences de l'Homme, 1983.

Gregory, Frederick. *Scientific Materialism in Nineteenth Century Germany.* Dordrecht: D. Reidel, 1977.

Gregory, Frederick. "Kant, Schelling, and the administration of science in the Romantic era." *Osiris* 2nd series, 5 (1989): 17–35.

Grell, Karl G. "Die Gastraea-Theorie." *Medizinhistorisches Journal* 14 (1979): 275–291.

Hall, Thomas S. "On biological analogs of Newtonian paradigms." *Philosophy of Science 35* (1968): 6–27.

Harrington, Anne. *Reenchanted Science: Holism in German Culture from Wilhelm II to Hitler.* Princeton: Princeton University Press, 1996.

Harwood, Jonathan. *Styles of Scientific Thought: The German Genetics Community, 1900–1933.* Chicago: University of Chicago Press, 1993.

Herbert, Sandra. *Charles Darwin, Geologist.* Ithaca: Cornell University Press, 2005.

Hodge, M. J. S. "Darwin's argument in the *Origin.*" *Philosophy of Science 59* (1992): 461–464.

Hopwood, Nick. "'Giving body' to embryos: Modeling, mechanism, and the microtome in late nineteenth-century anatomy." *Isis* 90 (1999): 462–496.

Hull, David L. "Exemplars and scientific change." *PSA—Proceedings of the Biennial Meeting of the Philosophy of Science Association* (1982), 2: 479–503.

Hull, David L. "Darwinism as a historical entity." In David Kohn, ed., *The Darwinian Heritage,* 773–812. Princeton: Princeton University Press, 1985.

Huxley, Julian S. *Evolution: The Modern Synthesis.* New York and London: Harper, 1942.

Jespersen, P. Helvig. "Charles Darwin and Dr. Grant." *Lychnos* (1948–1949): 159–167.

Johns, Adrian. *The Nature of the Book: Print and Knowledge in the Making.* Chicago: University of Chicago Press, 1998.

Junker, Thomas. "Heinrich Georg Bronn und die *Entstehung der Arten.*" *Sudhoffs Archiv für Geschichte der Medizin und der Naturwissenschaften* 75 (1991): 180–208.

Kanz, Kai Torsten. "Carl Friedrich Kielmeyer (1765–1844)—Leben, Werk, Wirkung: Perspektiven der Forschung und Edition." In Kai Torsten Kanz, ed., *Philosophie des Organischen in der Goethezeit: Studien zu Werk und Wirkung des Naturforschers Carl Friedrich Kielmeyer (1765–1844),* 13–32. Stuttgart: Franz Steiner, 1994.

Kanz, Kai Torsten, ed. *Philosophie des Organischen in der Goethezeit. Studien zu Werk und Wirkung des Naturforschers Carl Friedrich Kielmeyer (1765–1844).* Stuttgart: Franz Steiner, 1994.

Kanz, Kai Torsten. "'. . . die Biologie als die Krone oder der höchste Strebepunct aller Wissenschaften': Zur Rezeption des Biologiebegriffs in der romantischen Naturforschung (Lorenz Oken, Ernst Bartels, Carl Gustav Carus)." *NTM—Internationale Zeitschrift für Geschichte und Ethik der Naturwissenschaften, Technik und Medizin* 15 (2006): 77–92.

Kleeberg, Bernhard. *Theophysis: Ernst Haeckels Philosophie des Naturganzen.* Cologne: Böhlau, 2005.

Krauße, Erika. *Ernst Haeckel.* Leipzig: BSB B. G. Teubner, 1984.

Krauße, Erika. "Zum Verhältnis von Carl Gegenbaur (1826–1903) und Ernst Haeckel (1834–1919)." In Armin Geus, Wolfgang Gutmann, and Michael Weingarten, eds., *Miscellen zur Geschichte der Biologie: Ilse Jahn und Hans Querner zum 70. Geburtstag,* 83–89. Frankfurt: Waldemar Kramer, 1994.

Larson, James L. *Interpreting Nature: The Science of Living Form From Linnaeus to Kant.* Baltimore: Johns Hopkins University Press, 1994.

Lenoir, Timothy. "Kant, Blumenbach, and vital materialism in German biology." *Isis* 71 (1980): 77–108.

Lenoir, Timothy. "The Göttingen school and the development of transcendental *Naturphilosophie* in the romantic era." *Studies in History of Biology* 5 (1981): 111–205.

Lenoir, Timothy. *The Strategy of Life. Teleology and Mechanics in Nineteenth Century German Biology.* Dordrecht: D. Reidel, 1982.

Levit, Georgy S., and Uwe Hoßfeld. "The forgotten 'old-Darwinian' synthesis: The evolutionary theory of Ludwig H. Plate (1862–1937)." *NTM—Internationale Zeitschrift für Geschichte und Ethik der Naturwissenschaften, Technik und Medizin* 14 (2006): 9–25.

Livingstone, David N. *Putting Science in Its Place: Geographies of Scientific Knowledge.* Chicago: University of Chicago Press, 2003.

Lovejoy, Arthur O. *The Great Chain of Being: A Study in the History of an Idea.* Cambridge, Mass.: Harvard University Press, 1936.

Maienschein, Jane. "Arguments for experimentation in biology." *PSA—Proceedings of the Biennial Meeting of the Philosophy of Science Association* 2 (1986): 180–195.

Maienschein, Jane. "Shifting assumptions in American biology: Embryology, 1890–1910." *Journal of the History of Biology* 14 (1981): 89–113.

Mayr, Ernst. *The Growth of Biological Thought: Diversity, Evolution, and Inheritance.* Cambridge, Mass.: Harvard University/Belknap Press, 1982.

McLaughlin, Peter. "Blumenbach und der Bildungstrieb: Zum Verhältnis von epigenetischer Embryologie und typologischem Artbegriff." *Medizinhistorisches Journal* 17 (1982): 357–372.

McLaughlin, Peter. *Kant's Critique of Teleology in Biological Explanation: Antinomy and Teleology.* Lewiston: E. Mellen Press, 1990.

McLaughlin, Peter. "Naming biology." *Journal of the History of Biology* 35 (2002): 1–4.

McOuat, Gordon R. "Species, rules and meaning: The politics of language and the ends of definitions in 19th century natural history." *Studies in the History and Philosophy of Science* 27 (1996): 473–519.

Montgomery, Scott L. *Science in Translation: Movements of Knowledge Through Cultures and Time.* Chicago: University of Chicago Press, 2000.

Montgomery, William M. "Evolution and Darwinism in German biology, 1800–1883." Dissertation, University of Texas at Austin, 1974.

Montgomery, William M. "Germany." In Thomas F. Glick, ed., *The Comparative Reception of Darwinism*, 81–116. Austin: University of Texas Press, 1974.

Mylott, Anne. "The roots of cell theory in sap, spores, and Schleiden." Dissertation, Indiana University, 2002.

Nicolson, Malcolm. "Humboldtian plant geography after Humboldt: The link to ecology." *British Journal for the History of Science* 29 (1996): 289–310.

Nordenskiöld, Erik. *The History of Biology: A Survey*. Trans. Leonard Eyre. 1928. Reprint, New York: Tudor Publishing, 1936.

Nyhart, Lynn K. *Biology Takes Form: Animal Morphology and the German Universities, 1800–1900*. Chicago: University of Chicago Press, 1995.

Nyhart, Lynn K. "Civic and economic zoology in nineteenth-century Germany: The 'living communities' of Karl Möbius." *Isis* 89 (1999): 605–630.

Olesko, Kathryn M. "Introduction [to volume on 'Science in Germany']." *Osiris* 2nd series, 5 (1989): 7–14.

Ospovat, Dov. *The Development of Darwin's Theory: Natural History, Natural Theology, and Natural Selection, 1838–1859*. Cambridge: Cambridge University Press, 1981. Reprint, 1995.

Pancaldi, Giuliano. *Darwin in Italy: Science Across Cultural Frontiers*. 2nd ed. Trans. Ruey Morelli. Bloomington and Indianapolis: Indiana University Press, 1991.

Paulsen, Friedrich. *Die deutschen Universitäten und das Universitätsstudium*. Berlin: A. Ascher, 1902.

Pfannenstiel, Max. *Lorenz Oken, sein Leben und Wirken*. Freiburg im Breisgau: Hans Ferdinand Schulz, 1953.

Provine, William B. *The Origins of Theoretical Population Genetics*. Chicago: University of Chicago Press, 1971.

Provine, William B. "Progress in evolution and meaning in life." In C. Kenneth Waters and Albert van Helden, eds., *Julian Huxley: Biologist and Statesman of Science*, 165–180. Houston: Rice University Press, 1992.

Pusey, James R. *China and Charles Darwin*. Cambridge, Mass.: Harvard University Press, 1983.

Raikov, Boris. *Christian Heinrich Pander: Ein bedeutender Biologe und Evolutionist/An Important Biologist and Evolutionist, 1794–1865*. Trans. Woldemar von Hertzenberg and Peter von Bitter. Frankfurt am Main: Waldemar Kramer, 1984.

Rasmussen, Nicolas. "The decline of recapitulationism in early twentieth-century biology: Disciplinary conflict and consensus on the battleground of theory." *Journal of the History of Biology* 24 (1991): 51–89.

Recker, Doren A. "Causal efficacy: The structure of Darwin's argument strategy in the *Origin of Species*." *Philosophy of Science* 54 (1987): 147–175.

Rehbock, Philip F. *The Philosophical Naturalists: Themes in Nineteenth-Century British Biology*. Madison: University of Wisconsin Press, 1983.

Richards, Robert J. *The Meaning of Evolution: The Morphological Construction and Ideological Reconstruction of Darwin's Theory*. Chicago: University of Chicago Press, 1992.

Richards, Robert J. "Ideology and the history of science." *Biology and Philosophy* 8 (1993): 103–108.

Richards, Robert J. "The theological foundations of Darwin's theory of evolution." In Paul Theerman and Karen Parshall, eds., *Experiencing Nature: Proceedings of a Conference in Honor of Allen G. Debus*, 61–79. Dordrecht: Kluwer Academic, 1997.

Richards, Robert J. "Kant and Blumenbach on the *Bildungstrieb*: A historical misunderstanding." *Studies in the History and Philosophy of Biological and Biomedical Sciences* 31 (2000): 11–32.

Richards, Robert J. *The Romantic Conception of Life: Science and Philosophy in the Age of Goethe*. Chicago: University of Chicago Press, 2002.

Richards, Robert J. "If this be heresy: Haeckel's conversion to Darwinism." In Abigail Lustig, Robert J. Richards, and Michael Ruse, eds., *Darwinian Heresies*, 101–130. Cambridge and New York: Cambridge University Press, 2004.

Roe, Shirley A. *Matter, Life, and Generation: Eighteenth-century Embryology and the Haller–Wolff Debate*. Cambridge and New York: Cambridge University Press, 1981.

Rupke, Nicolaas A. "Richard Owen's vertebrate archetype." *Isis* 84 (1993): 231–251.

Rupke, Nicolaas A. *Richard Owen: Victorian Naturalist*. New Haven: Yale University Press, 1994.

Rupke, Nicolaas A. "Humboldtian medicine." *Medical History* 40 (1996): 292–310.

Rupke, Nicolaas A. "Translation studies in the history of science: The example of *Vestiges*." *British Journal for the History of Science* 33 (2000): 209–222.

Ruse, Michael. *The Darwinian Revolution: Science Red in Tooth and Claw*. Chicago: University of Chicago Press, 1979.

Russell, E. S. *Form and Function: A Contribution to the History of Animal Morphology*. London: John Murray, 1916.

Schilling, Dietmar. "Biographische und problemgeschichtliche Einleitung." Introduction to Jean-Baptiste Lamarck, *Zoologische Philosophie*. Leipzig: Akademische Verlagsgesellschaft Geest und Portig, 1990.

Schleiden, Matthias. "Methodologische Einleitung [from the 4th ed. of *Grundzüge der Wissenschaftlichen Botanik* (1861)]." In Ulrich Charpa, ed., *Wissenschaftsphilosophische Schriften*, 47–196. Cologne: Jürgen Dinter, Verlag für Philosophie, 1989.

Schmid, Günther. "Über die Herkunft der Ausdrücke Morphologie und Biologie." *Nova Acta Leopoldina* NF 2 (1935): 597–620.

Schumacher, Ingrid. "Die Entwicklungstheorie des Heidelberger Paläontologen und Zoologen Heinrich Georg Bronn (1800–1862)." Dissertation, Ruprecht-Karl-Universität, Heidelberg, 1975.

Schweber, Sylvan S. "The wider context in Darwin's theorizing." In David Kohn, ed., *The Darwinian Heritage*, 35–69. Princeton: Princeton University Press, 1985.

Secord, James A. "Nature's fancy: Charles Darwin and the breeding of pigeons." *Isis* 72 (1981): 162–186.

Secord, James A. "Darwin and the breeders: A social history." In David Kohn, ed., *The Darwinian Heritage*, 519–542. Princeton: Princeton University Press, 1985.

Secord, James A. *Victorian Sensation: The Extraordinary Publication, Reception, and Secret Authorship of* Vestiges of the Natural History of Creation. Chicago: University of Chicago Press, 2000.

Secord, James A. "Knowledge in transit." *Isis* 95 (2005): 654–672.

Sengoopta, Chandak. "Glandular politics: Experimental biology, clinical medicine, and homosexual emancipation in fin-de-siècle central Europe." *Isis* 89 (1998): 445–473.

Shaffer, Elinor S. "Romantic philosophy and the organization of the disciplines: The founding of the Humboldt University of Berlin." In Andrew Cunningham and Nicholas Jardine, eds., *Romanticism and the Sciences*, 38–54. Cambridge: Cambridge University Press, 1990.

Shanahan, Timothy. *The Evolution of Darwinism: Selection, Adaptation, and Progress in Evolutionary Biology*. Cambridge and New York: Cambridge University Press, 2004.

Sloan, Phillip R. "Darwin's invertebrate program, 1826–1836: Preconditions for transformism." In David Kohn, ed., *The Darwinian Heritage*, 71–120. Princeton: Princeton University Press, 1985.

Sloan, Phillip R. "Darwin, vital matter, and the transformation of species." *Journal of the History of Biology* 19 (1986): 369–445.

Sloan, Phillip R. "Introductory essay: On the edge of evolution." *The Hunterian Lectures in Comparative Anatomy, May-June, 1837*, 3–72. Chicago: University of Chicago Press, 1992.

Sober, Elliott. *The Nature of Selection*. Cambridge, Mass.: MIT Press, 1982.

Stott, Rebecca. *Darwin and the Barnacle: The Story of One Tiny Creature and History's Most Spectacular Scientific Breakthrough*. London: Faber and Faber, 2003.

Strick, James. *Sparks of Life: Darwinism and the Victorian Debates Over Spontaneous Generation*. Cambridge, Mass.: Harvard University Press, 2000.

Thomas, R. Hinton. *Liberalism, Nationalism, and the German Intellectuals (1822–1847): An Analysis of the Academic and Scientific Conferences of the Period*. Cambridge: W. Heffer, 1951.

Todes, Daniel P. *Darwin without Malthus: The Struggle for Existence in Russian Evolutionary Thought.* New York and Oxford: Oxford University Press, 1989.

Tuchman, Arleen. "Institutions and disciplines: Recent work in the history of German science." *Journal of Modern History* 69 (1997): 298–319.

Tuchman, Arleen. *Science, Medicine and the State in Germany: The Case of Baden, 1815–1871.* New York and Oxford: Oxford University Press, 1994.

Turner, R. Steven. "The growth of professorial research in Prussia, 1818 to 1848—causes and context." *Historical Studies in the Physical Sciences* 3 (1971): 137–182.

Turner, R. Steven. "The Prussian Universities and the Research Imperative." Dissertation, Princeton University, 1973.

Uschmann, Georg. *Der morphologische Vervollkommnungsbegriff bei Goethe und seine problemgeschichtlichen Zusammenhänge.* Jena: Gustav Fischer, 1939.

Uschmann, Georg. *Geschichte der Zoologie und der zoologischen Anstalten in Jena 1771–1919.* Jena: VEB Gustav Fischer Verlag, 1959.

Weikart, Richard. *From Darwin to Hitler: Evolutionary Ethics, Eugenics, and Racism in Germany.* New York: Palgrave Macmillan, 2004.

Weikart, Richard. "Re: REV: Gliboff on Weikart, _Darwin to Hitler_ (Weikart)." Electronic publication: H-German, 2004, from http://h-net.msu.edu/cgi-bin/logbrowse.pl?trx=lx&list=H-German&user=&pw=&month=0409.

Weikart, Richard. *Socialist Darwinism: Evolution in German Socialist Thought from Marx to Bernstein.* San Francisco: International Scholars Publications, 1999.

Weiling, Franz. "J. G. Mendel sowie die von M. Pettenkofer angeregten Untersuchungen des Zusammenhanges von Cholera- und Typhus-Massenerkrankungen mit dem Grundwasserstand." *Sudhoffs Archiv für Geschichte der Medizin und der Naturwissenschaften* 59 (1975): 1–19.

Weindling, Paul. "Ernst Haeckel, Darwinismus and the secularization of nature." In James R. Moore, ed., *History, Humanity, and Evolution: Essays for John C. Greene*, 311–327. Cambridge: Cambridge University Press, 1989.

Winsor, Mary P. *Starfish, Jellyfish, and the Order of Life.* New Haven: Yale University Press, 1976.

Winsor, Mary P. "Non-essentialist methods in pre-Darwinian taxonomy." *Biology and Philosophy* 18 (2003): 387–400.

Winther, Rasmus G. "August Weismann on germ-plasm variation." *Journal of the History of Biology* 34 (2001): 517–555.

Young, Robert M. "Darwin's metaphor: Does nature select?" In *Darwin's Metaphor: Nature's Place in Victorian Culture*, 79–125. Cambridge: Cambridge University Press, 1985.

Young, Robert M. "Malthus and the evolutionists: The common context of biological and social theory." In *Darwin's Metaphor: Nature's Place in Victorian Culture*, 23–55. Cambridge: Cambridge University Press, 1985.

Index